Matrix Methods of Structural Analysis

Matrix Methods of Structural Analysis

Praveen Nagarajan

CRC Press
Taylor & Francis Group
Boca Raton London New York

CRC Press is an imprint of the
Taylor & Francis Group, an **informa** business

CRC Press
Taylor & Francis Group
6000 Broken Sound Parkway NW, Suite 300
Boca Raton, FL 33487-2742

First issued in paperback 2020

© 2019 by Taylor & Francis Group, LLC
CRC Press is an imprint of Taylor & Francis Group, an Informa business

No claim to original U.S. Government works

ISBN 13: 978-0-367-57126-9 (pbk)
ISBN 13: 978-0-8153-8150-1 (hbk)

Library of Congress Cataloging-in-Publication Data

Names: Nagarajan, Praveen, author.
Title: Matrix methods of structural analysis / Praveen Nagarajan.
Description: Boca Raton : Taylor & Francis, a CRC title, part of the Taylor & Francis imprint, a member of the Taylor & Francis Group, the academic division of T&F Informa, plc, 2019. | Includes bibliographical references and index.
Identifiers: LCCN 2018016659 | ISBN 9780815381501 (hardback : acid-free paper) | ISBN 9781351210324 (ebook)
Subjects: LCSH: Structural analysis (Engineering)—Matrix methods.
Classification: LCC TA642 .N34 2019 | DDC 624.1/7015129434—dc23
LC record available at https://lccn.loc.gov/2018016659

**Visit the Taylor & Francis Web site at
http://www.taylorandfrancis.com**

**and the CRC Press Web site at
http://www.crcpress.com**

To my parents

Mrs. Rani Nagarajan and Dr. N.M. Nagarajan

and my Teachers at

NIT Calicut and IIT Madras

Contents

Preface

Matrix methods of structural analysis are used for the analysis of the framed structures, i.e., structures composed of one-dimensional elements. The solution procedures used in this method are systematic and general. Hence, it is easier to write computer programs using this method.

The matrix method of analysis is an important topic in the field of structural engineering. A proper knowledge in this area is necessary to understand advance topics like finite element method, structural dynamics, structural stability, etc. The software available for the analysis of structures is developed using this method. Hence, it is necessary to have a background in this subject for the proper application of these software tools.

This book deals with analysis of structures using matrix methods. It is designed as an easy-to-read textbook for an introductory course in matrix methods of structural analysis for senior undergraduate and postgraduate students of civil engineering. It will be also useful for practicing structural engineers for the efficient usage of structural analysis software tools.

The key features of the book are as follows:

- The concepts are explained in a simple manner.
- Each chapter contains plenty of illustrations and examples for improved understanding.
- A step-by-step approach for problem solving is adopted in the book which will help the reader to solve problems in a systematic way.

The book contains five chapters. In the first chapter, the background to matrix analysis of structures is discussed. The procedure to develop force–displacement relation for a given structure using flexibility and stiffness coefficients is covered in the second chapter. The remaining three chapters deal with the analysis of framed structures using the flexibility, stiffness, and direct stiffness method. A simple MATLAB code for analyzing plane truss using the direct stiffness method is given in the appendix of the book. The code can be modified so that it can be used for the analysis of other types of structures.

The author sincerely thanks Dr. Gagandeep Singh , Ms. Mouli Sharma and Mr. Edward Curtis of CRC Press and Ms. Alexandra Andrejevich of codeMantra for all the support and to Mr. Beljith P., for typing the manuscript.

The author welcomes comments, feedback, and suggestions for the improvement in the future edition of the book.

Praveen Nagarajan

MATLAB® is a registered trademark of The MathWorks, Inc. For product information, please contact:

The MathWorks, Inc.
3 Apple Hill Drive
Natick, MA 01760-2098 USA
Tel: 508-647-7000
Fax: 508-647-7001
E-mail: info@mathworks.com
Web: www.mathworks.com

Author

Praveen Nagarajan completed his civil engineering education from NIT Calicut and IIT Madras. He was a top-ranking student throughout. After a brief stint as Bridge design Engineer at L&T Ramboll, Chennai, he took to academics.

His research interests include reinforced and prestressed concrete, bridge engineering, and structural reliability. He has published more than 50 technical papers in these areas and authored the book *Prestressed Concrete Design* (published by Pearson). He is the recipient of several awards like the Valli Anantharamakrishnan Merit Prize from IIT Madras, E P Nicolaides Prize from the Institution of Engineers (India), the Best Young Teacher award from NIT Calicut, ICI-UltraTech Award for Outstanding Young Concrete Engineer of Kerala by the Indian Concrete Institute (ICI), and ICI-Prof. V. Ramakrishnan Young Scientist Award by the ICI. He has guided two PhD students and more than 35 M-Tech projects. He is also guiding eight research scholars for their doctoral degrees. Currently, he is working as an Associate Professor in the Department of Civil Engineering at National Institute of Technology Calicut, India.

1

Introduction

1.1 Introduction to Matrix Methods of Structural Analysis

Structural analysis deals with the determination of response (forces and displacements) of the structure subjected to loads. The rapid development of computers and the need for complex and light weight structures lead to the development of matrix methods of structural analysis. The analysis procedure can be concisely written using matrix notations and are suitable for computer programming.

The classical methods of structural analysis (also known as *system approach of analysis*), such as the method of consistent deformation, slope-deflection methods, etc., consider the behavior of the entire structure for developing equations necessary for analysis. Hence, these methods are suitable only for simple or small structures. For larger structures, the use of classical methods of analysis will be difficult and time-consuming. Even though the procedures using classical methods can be written in matrix format, the matrix methods of analysis discussed in this book is different from the classical approach and is known as the *element approach of analysis*. In this method, the response of the whole structure is determined using the behavior of the elements or members from which the structure is made of.

There exists two ways by which a structure is analyzed using matrix methods, namely the *flexibility method* or the *force method* and the *stiffness method* or the *displacement method*. The flexibility method of analysis is discussed in Chapter 3, whereas Chapters 4 and 5 deal with the stiffness method of analysis. The methods to develop the force–displacement relation of elements are covered in Chapter 2. In this chapter, the important terms and their definitions used in matrix methods of analysis are discussed.

1.2 Framed Structures

A structure is defined as an assemblage of structural elements that can withstand load. When two dimensions (cross-sectional dimensions) of the elements are very small compared to the third dimension (length of element), the elements are modeled as one-dimensional *line elements*. Elements having one dimension (thickness) very small compared to the other two dimensions are modeled as two-dimensional *surface elements*. In some cases, all the dimensions of the element are of comparable length. These elements are modeled as three-dimensional *brick elements*.

A structure composed of only line elements is known as a *framed structure* or *skeletal structure*. This book deals with the analysis of framed structures. In a framed structure, joints represent the point of intersection of elements, supports, and free end of elements (A, B, C, D and E in Figure 1.1a). The supports can be fixed (support A in Figure 1.1a and 1.1b), hinged (support D in Figure 1.1a), roller (support B in Figure 1.1b), or guided-fixed (support C in Figure 1.1b). Sometimes supports can be elastic (or spring). The framed structures can be grouped into six categories, namely (i) beam, (ii) plane truss, (iii) space truss, (iv) plane frame, (v) grid, and (vi) space frame. A brief description of each type of framed structure is given below:

A beam (Figure 1.2a) is a long-slender element having one or more points of support. The forces act in a plane containing an axis of symmetry of the cross-section of the element and the direction of moment vector will be perpendicular to this plane. The beam also deflects in this plane. At any section of the beam, the internal forces developed are the bending moment and shear force (Figure 1.3b). It is assumed that the loads are transverse to the axis of element; hence, axial force is not developed.

Plane truss (Figure 1.2b) consists of line elements, all lying in a plane and interconnected by frictionless hinges. The loads and reactions act only at these hinges. Thus, the elements develop only axial forces (Figure 1.3a).

A space truss (Figure 1.2c) is similar to a plane truss. However, the elements, loads, and reactions can have any orientation in space. Also in this case, only axial forces are developed in the element (Figure 1.3a).

In the case of a plane frame (Figure 1.2d), all the elements lie in a single plane and, similar to the beam, the axes of symmetry of the cross-section also lie in the plane. The force vector and displacement of the structure occur in this plane. The moment vector is perpendicular to this plane. In a plane frame element, at any section, axial force, shear force, and bending moment are developed (Figure 1.3c).

A grid (Figure 1.2e) is similar to plane frame, but with a difference. In this case, all members lie in a plane but the direction of force vector is perpendicular to this plane. The moment vector lies in the plane of members. The internal forces developed in a grid element are shear force, bending moment, and twisting moment (Figure 1.3d).

A space frame (Figure 1.2f) is the most general type of framed structure. In this case, the elements and forces can have any orientation in space. The internal forces developed at any section of the space frame element are axial force, twisting moment, bending moment, and shear force in both the principal directions of the cross-section of the element (Figure 1.3e).

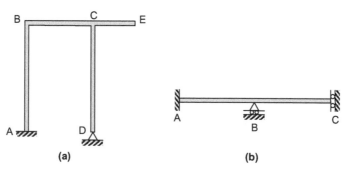

(a)   (b)

FIGURE 1.1
Joints in a framed structure.

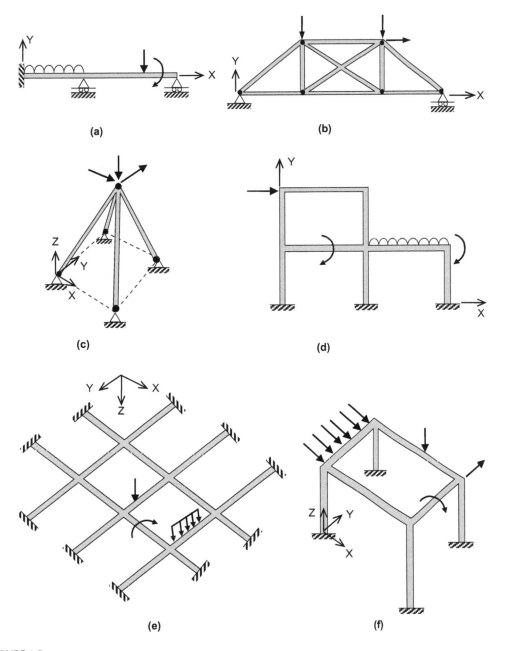

FIGURE 1.2
Types of framed structure: (a) beam, (b) plane truss, (c) space truss, (d) plane frame, (e) grid, and (f) space frame.

Figure 1.3 shows the Free Body Diagram (FBD) of different types of elements in a framed structure indicating the nature of internal forces developed in the element. In this figure, N, M, V, and T are the axial force, bending moment, shear force, and twisting moment, respectively. The arrow with single head and double head shows the direction of force vector and moment vector, respectively.

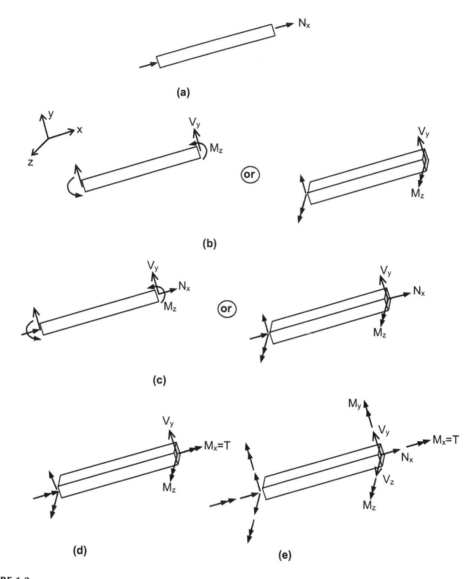

FIGURE 1.3
Internal forces in different types of line elements: (a) truss element, (b) beam element, (c) plane frame element, (d) grid element, and (e) space frame element.

1.3 Cartesian Coordinate System

A right-handed orthogonal Cartesian coordinate system as shown in Figure 1.4a is used in this book to specify the geometry, forces, and displacements in a structure. The three orthogonal axes X, Y, and Z with their positive directions are indicated in the figure. Forces (the term forces also includes moments) and displacements (translations and rotations) are taken as positive, if they are along the positive directions of the coordinate system. For example, in the case of a two-dimensional structure, the positive direction of distances (x, y), displacements (u, v, θ_z), and forces (P_x, P_y, M_z) are shown in Figure 1.4b.

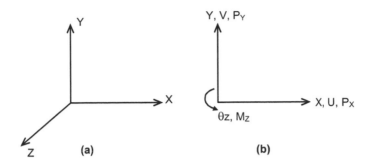

FIGURE 1.4
Right-handed orthogonal Cartesian coordinate system.

The positive directions for the forces and displacements at the two ends of the elements are shown in Figure 1.3.

1.4 Coordinate Systems for Forces and Displacements

In order to define a force or displacement, it is necessary to specify (i) the location on the structure where the force is applied or displacement is measured, (ii) the magnitude, and (iii) direction. Forces (or displacements) can be easily defined by marking them on a sketch of the structure. Figure 1.5a shows one such way of indicating forces in a plane frame structure.

Consider a situation in which the plane frame is subjected to different combinations of forces but the locations and directions of forces are the same as shown in Figure 1.5a. In such cases, it is convenient to define a coordinate system as shown in Figure 1.5b that indicates the points of application and directions of forces and the corresponding displacements (similarly if a coordinate system is developed to indicate displacements, then they

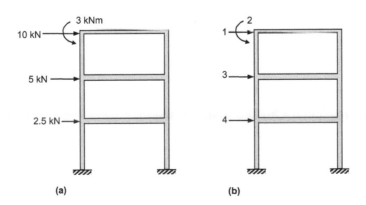

FIGURE 1.5
Coordinate system for a structure: (a) force acting on structure and (b) coordinates for forces and displacements.

also represent the direction of corresponding forces). The magnitude of forces acting along the coordinates are indicated in a column matrix as

$$\left\{ \begin{array}{c} P_1 \\ P_2 \\ P_3 \\ P_4 \end{array} \right\}$$

This column matrix (or vector) is known as the *force vector*. A negative quantity in the force vector only indicates that the force is acting in a direction opposite to the direction shown in the coordinate system. The corresponding displacements are also represented in a column matrix as

$$\left\{ \begin{array}{c} D_1 \\ D_2 \\ D_3 \\ D_4 \end{array} \right\}$$

This matrix is known as the *displacement vector*. The force and displacement vectors are written in a compact form as {P} and {D}, respectively.*

1.5 Nodes and Elements

The structure is considered as an assembly of elements. The elements are connected to one other at a finite number of points known as *nodes*. The joints in a structure are also a type of node. In a framed structure, each element has only two nodes and the different types of elements are already discussed in Section 1.2. Also it should be noted that in matrix methods of structural analysis, the forces and displacements can be applied or determined at nodes.

The discretized structure indicating the nodes and elements of a typical truss, beam, and plane frame is shown in Figure 1.6. In the figure, elements are specified by a number enclosed in a square. Similarly, the nodes are indicated by a number enclosed in a circle.

1.6 Nodal Degrees of Freedom

The degrees of freedom (DOF) of a node are the number of independent displacements that the node can undergo. The DOF of a node in different types of framed structures are given in Table 1.1.

The nodal DOF of the structures shown in Figure 1.6 are indicated in Figure 1.7. In the figure, translational and rotational DOF are shown by straight and curved arrows, respectively.

---

* The flower braces {} indicate that the matrix is a column matrix.

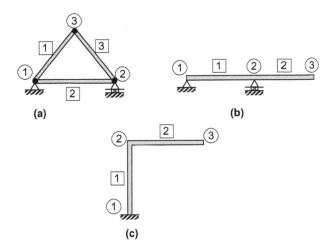

FIGURE 1.6
Nodes and elements in typical framed structures: (a) plane truss, (b) beam, and (c) plane frame.

TABLE 1.1

Nodal DOF in a Framed Structure

No.	Type of Structure	DOF per Node	Description
1.	Plane truss	2	All translational
2.	Beam	2	One translational and one rotational
3.	Plane frame	3	Two translational and one rotational
4.	Grid	3	One translational and two rotational
5.	Space truss	3	All translational
6.	Space frame	6	Three translational and three rotational

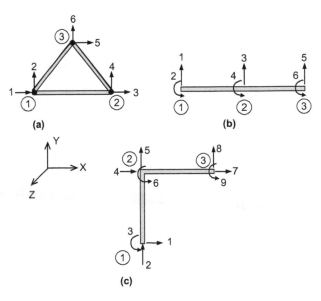

FIGURE 1.7
Nodal DOF of typical framed structures: (a) plane truss, (b) beam, and (c) plane frame.

From the figure, it can be seen that each DOF is identified by a number. The numbering of the DOF starts from the node having lower most number and the proceeded to the next node and so on. At a node, the translational DOF is numbered before rotational DOF. If there is more than one translational (or rotational) DOF at a node, then DOF in the direction of X-axis is numbered first followed by DOF along the Y-axis and Z-axis.

1.7 Global and Local Coordinate System

For the analysis of structures, two different coordinate systems for forces and displacements are used. A single *global coordinate system* (X, Y, and Z are corresponding *global axes*) is used to specify the forces (P) and displacements (D) at the nodes of the structure. A *local coordinate system* (x, y, and z are the corresponding *local axes*) will be used for each element to indicate the element forces (p) and the corresponding displacements (d). The global coordinates of structures discussed in Section 1.5 are shown in Figure 1.8. The local coordinates of one typical element of the structure are also shown. In the figure, x, y, and z indicate the local axes.

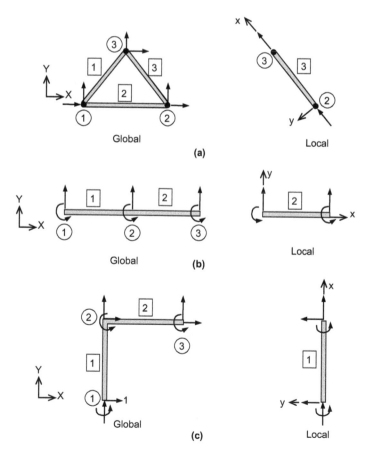

FIGURE 1.8
Global and local coordinates: (a) plane truss, (b) beam, and (c) plane frame.

All the elements in a framed structure will have two nodes. To distinguish the two nodes in an element: one node is called as the *near node* and the other as the *far node*. The node with the smallest node number is usually taken as the near node. Also, the near node is treated as the origin of the local axes. The direction from the near node to the far node is chosen as the x-axis. The principal axes of the cross-section of the element will be the y- and z-axes. In this way, it will be easy to specify the cross-sectional properties (say the moment of inertia of the section) along the local axes. Furthermore, from the design point of view, it is convenient to obtain the element forces with respect to the local axes.

1.8 Specification of Geometry of the Structure

The geometry of the structure is developed using the following details:

a. Nodal coordinate data—From this, the coordinates of all the nodes are obtained.

b. Element connectivity data—This list gives the details of the elements and the nodes to which elements are connected.

The nodal coordinates and element connectivity data for the truss shown in Figure 1.9 are given in Tables 1.2 and 1.3.

Using the information given in Tables 1.2 and 1.3, the geometry of the truss is developed.

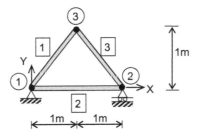

FIGURE 1.9
Nodes and elements of a truss.

TABLE 1.2

Nodal Coordinates Data of the Truss

Nodal Number	Coordinates (m)		
	X	Y	Z
1	0	0	0
2	2	0	0
3	1	1	0

TABLE 1.3

Element Connectivity Data of the Truss

Element Number	Near Node Number	Far Node Number
1	1	3
2	1	2
3	2	3

1.9 Equivalent Nodal Loads

In the matrix method of analysis, it is assumed that the loads act only at the nodes. However, a structure may be subjected to distributed loads or other types of loads acting on the elements. These loads are replaced by equivalent loads acting at the nodes. The loads acting on the structure are grouped into two categories, namely the loads acting on the nodes known as *nodal loads P* and the loads acting on the elements known as *element loads*. The element loads are replaced by *equivalent nodal loads* P_e. The equivalent nodal loads are added to the real nodal loads P. The total load thus obtained is known as the *combined nodal loads* P_c ($P_c = P + P_e$). The structure subjected to combined nodal loads is then analyzed using matrix methods. The principle used to find the equivalent nodal load P_e is that the nodal displacements obtained by analyzing the structure subject to combined nodal loads P_c will be equal to displacement of the structure caused by actual loads.

The procedure for finding P_e and P_c is explained using a two-span continuous beam subjected to nodal and element loads (Figure 1.10a). The FDB indicating the loads and support reactions is depicted in Figure 1.10b. The discretized structure is shown in Figure 1.10c. Since in this case the structure is modeled using only two elements, the concentrated loads W_1 and W_2 are treated as element loads. The loads acting on the structure are grouped into nodal loads (Figure 1.10d) and element loads (Figure 1.10e), respectively.

Consider the structure subjected to element loads (Figure 1.10e). The structure is restrained against all nodal displacements (DOF = 0) by introducing nodal restraints (i.e., all the nodes of the structure are fixed) as shown in Figure 1.10f. The support reactions of the restrained structures are found out. Since all the nodes are fixed, the support reactions are also known as *fixed end actions* P_F. For clarity, the fixed elements are shown separately in Figure 1.10g and the fixed end actions are determined. The fixed end actions for typical loads and end displacements are shown in Figure 1.11. Figure 1.10h shows the fixed end actions of the restrained structure (in the figure, the reactions are indicated by arrows with slashed line).

The effects of artificially fixing the nodes are neutralized by reversing the fixed end actions and applying them on the structure (Figure 1.10i). The reversed fixed end actions are known as the *equivalent nodal loads* P_e ($P_e = -P_F$). The nodal loads (Figure 1.10d) are added with these equivalent nodal loads to get the combined nodal loads (Figure 1.10j). The structure is then analyzed for these loads.

Figure 1.10k shows the FBD of the structure subjected to combined loads. The superposition of this FBD with Figure 1.10h leads to the force system shown in Figure 1.10l. Comparing this with the force system in Figure 1.10b indicates that the original force system can be considered as the superposition of the force systems shown in Figures 1.10h and 1.10j.

Initially, it was mentioned that the nodal displacements due to combined loads will be the same as due to the real loads. To observe this, consider the superposition of the nodal displacements of the structures shown in Figures 1.10h and 1.10j. This gives the nodal displacements of the structure subjected to real loads (Figure 1.10a). However, in Figure 1.10h, since all the nodes are fixed, the nodal displacements in this case are zero. Hence, the nodal displacements of the structure subjected to combined loads will be equal to the nodal displacements caused by actual loads.

Similarly, it can be proved that the support reactions of the structures subjected to combined loads are equal to the support reactions of the structure subjected to real loads. For this, consider the superposition of the force systems of Figures 1.10h and 1.10k. The combined effect is shown in Figure 1.10l. Comparing this with FBD of the structure subjected to actual loads (Figure 1.10b) indicates that the support reactions due to combined nodal loads (A'_y, M'_A and B'_y) are the same as the reactions due to actual loads (A_y, M_A, and B_y).

However, the element forces obtained from the analysis of structure subjected to combined nodal loads are not equal to the element forces in the structure subjected to actual loads. The element forces are obtained by adding the element forces in the restrained structure subjected to element loads (Figure 1.10h) with the element forces in the structure loaded by combined loads (Figure 1.10g).

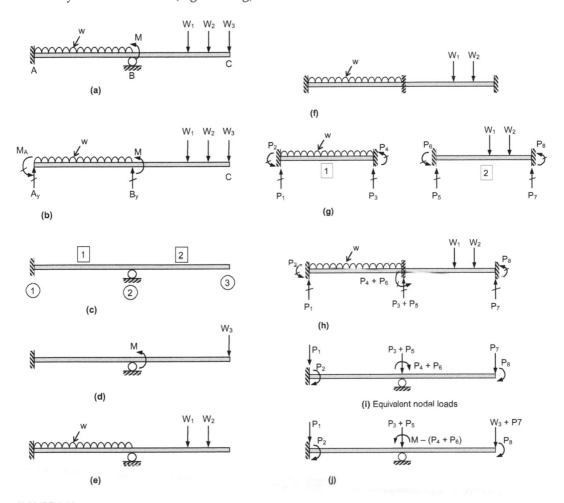

FIGURE 1.10
Equivalent and combined nodal loads: (a) structure with loads, (b) FBD of the structure subjected to actual loads, (c) discretized structure, (d) nodal loads, (e) element loads, (f) restrained structure subjected to element loads, (g) fixed end actions for elements 1 and 2, (h) fixed end actions for restrained structure, (i) equivalent nodal loads, (j) structure subjected to combined loads, (k) FBD of structure subjected to combined loads, and (l) addition of forces in Figures 1.10h and k.

(Continued)

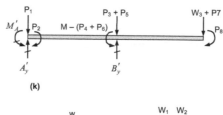

(k)

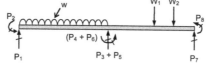

FBD of restrained structure (Fig. 1.10h)

⊕

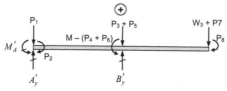

FBD of structure subjected to combined loads (Fig. 1.10k)

⊜

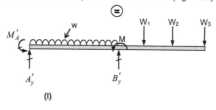

(l)

FIGURE 1.10 (CONTINUED)
Equivalent and combined nodal loads: (a) structure with loads, (b) FBD of the structure subjected to actual loads, (c) discretized structure, (d) nodal loads, (e) element loads, (f) restrained structure subjected to element loads, (g) fixed end actions for elements 1 and 2, (h) fixed end actions for restrained structure, (i) equivalent nodal loads, (j) structure subjected to combined loads, (k) FBD of structure subjected to combined loads, and (l) addition of forces in Figures 1.10h and k.

Example 1.1

Determine the nodal loads required for analyzing the continuous beam shown in Figure 1.12a.

Solution

Step 1: Develop the discretized structure
    The nodes and elements are shown in Figure 1.12b. Here, the 8 kN force is treated as an element load. Otherwise additional node should be included at the point of application of this force so that the load can be treated as a nodal load.
Step 2: Identify the nodal and element loads
    The nodal and elemental loads are shown in Figures 1.12c and 1.12d.
Step 3: Find the fixed end actions
    The restrained structure (all the nodes are fixed) subjected to element loads is shown in Figure 1.12e. For convenience, the fixed end actions of the individual elements are determined (Figure 1.12f) and the fixed end actions of the total structure are found out (Figure 1.12 g).
Step 4: Determine the equivalent nodal loads
    The equivalent nodal loads (Figure 1.12h) are obtained by reversing the fixed end actions.

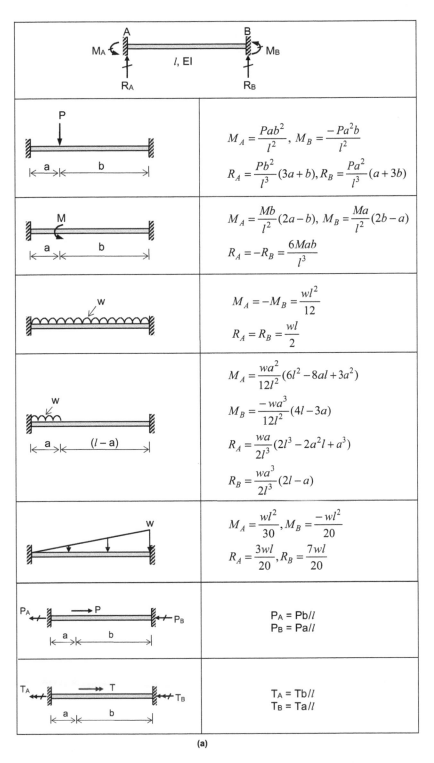

FIGURE 1.11
Fixed end actions: (a) due to loads and (b) due to displacements.

(*Continued*)

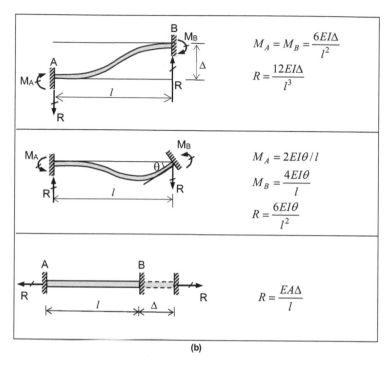

FIGURE 1.11 (CONTINUED)
Fixed end actions: (a) due to loads and (b) due to displacements.

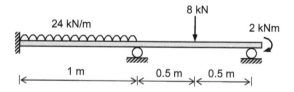

FIGURE 1.12a
Continuous beam.

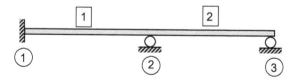

FIGURE 1.12b
Discretized structure.

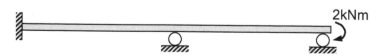

FIGURE 1.12c
Nodal loads.

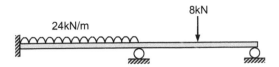

FIGURE 1.12d
Element loads.

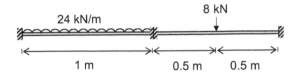

FIGURE 1.12e
Restrained structure.

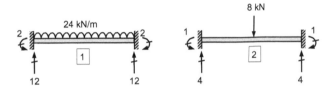

FIGURE 1.12f
Fixed end actions of individual elements.

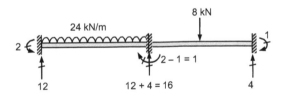

FIGURE 1.12g
Fixed end actions for the structure.

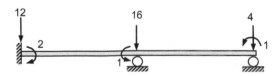

FIGURE 1.12h
Equivalent nodal loads.

Step 5: Determine the combined nodal loads

The combined nodal loads (loads required for analysis) are obtained by adding the nodal loads (Figure 1.12c) with the equivalent nodal loads (Figure 1.12h). The combined nodal loads for the structure are shown in Figure 1.12i.

Example 1.2

Determine the nodal loads required for analyzing the plane frame shown in Figure 1.13a.

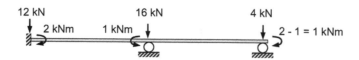

FIGURE 1.12i
Combined nodal loads.

Solution

Step 1: Develop the discretized structure.
   The elements and nodes are shown in Figure 1.13b.
Step 2: Identify the nodal and element loads
   The nodal and element loads are shown in Figures 1.13c and d.
Step 3: Find the fixed end actions
   The restrained structure subjected to element loads, fixed end actions of elements, and structures are shown in Figures 1.13e–g, respectively.
Step 4: Determine the equivalent nodal loads
   Figure 1.13h shows the equivalent nodal loads for the structure.
Step 5: Determine the combined nodal loads
   The combined nodal loads required for analysis are shown in Figure 1.13i.

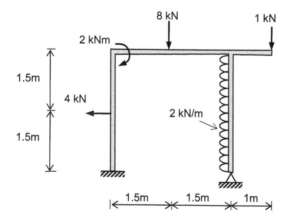

FIGURE 1.13a
Plane frame.

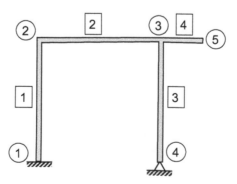

FIGURE 1.13b
Discretized structure.

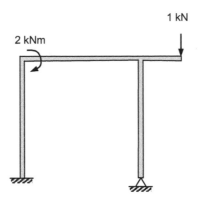

FIGURE 1.13c
Nodal loads.

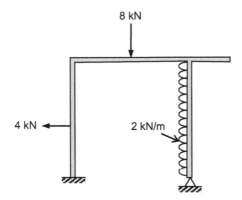

FIGURE 1.13d
Element loads.

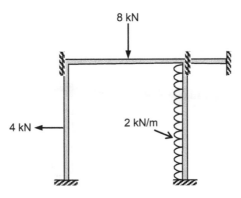

FIGURE 1.13e
Restrained structure subjected to element loads.

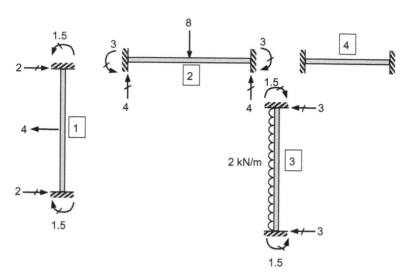

FIGURE 1.13f
Fixed end actions of individual elements.

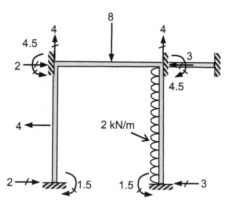

FIGURE 1.13g
Fixed end actions of the structure.

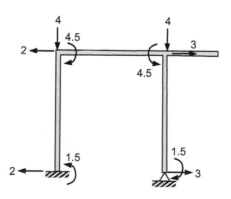

FIGURE 1.13h
Equivalent nodal loads.

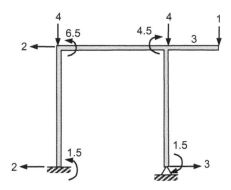

FIGURE 1.13i
Combined nodal loads.

1.10 Kinematic and Static Indeterminacy

Depending on the method of analysis, there are two types of indeterminacy, namely *degree of kinematic indeterminacy* (DKI) and *degree of static indeterminacy* (DSI). If the stiffness method (or displacement method) is used, then determination of DKI is necessary. When the flexibility method (or force method) is used for analysis, it becomes necessary to consider DSI.

1.10.1 Degree of Kinematic Indeterminacy (DKI)

The nodal DOF can be grouped into two, namely the *unconstrained* DOF and *constrained* DOF. The unconstrained DOF (also known as *active* DOF or *free* DOF) are the unknown nodal displacements or unknown nodal DOF. However, the nodal forces corresponding to these DOF are known. The nodal forces are the loads acting on the structure which are known. The constrained DOF or *restrained* DOF are the known DOF (they are the restrained DOF at the supports). But the corresponding nodal forces are unknown. The nodal forces are the support reactions which are not known.

In the stiffness method of analysis, the nodal displacements are the unknown quantities. Thus, in this method, the unconstrained DOF are the basic unknowns. The number of unconstrained DOF or active DOF of a structure is the DKI of the structure.

1.10.2 Degree of Static Indeterminacy (DSI)

In flexibility method, forces are treated as unknowns. A *statically determinate structure* is one in which all the unknown forces (support reactions and element forces) can be determined using equations of equilibrium alone. In some structures, the equations of equilibrium alone will not be sufficient to find the reactions and/or element forces. These types of structures are known as *statically indeterminate structure*. A structure can be statically indeterminate externally (in this case excess supports will be present) or statically indeterminate internally (due to excess elements) or both (due to excess supports and elements).

The DSI is equal to the number of equations in excess of equations of equilibrium necessary to find all the forces in the structure. It can be also found by finding the number of supports and/or elements that can be removed from the structure so that a statically determinate and stable structure is obtained.[*] Thus, we have

$$DSI = DSI_e + DSI_i \tag{1.1}$$

where DSI_e is the external indeterminacy (number of excess reaction components) and DSI_i is the internal indeterminacy (number of excess element forces).

For simple structures, DSI can be determined by inspection. In general, DSI is determined using the following equation:

$$DSI = N_u - N_E{}^{[\dagger]} \tag{1.2}$$

where N_u is the number of unknown forces (reactions and element forces) and N_E is the number of available equations (equations of equilibrium using the FBD of the nodes and *equations of condition*[‡]).

DSI_e is obtained by

$$DSI_e = N_R - N_{ES} \tag{1.3}$$

where N_R is the number of reactions and N_{ES} is the sum of equations of equilibrium for the structure (i.e., by considering the FBD of the complete structure) and the equations of condition.

Once DSI_e is determined, DSI_i is calculated using Eq. 1.1. Thus, we have

$$DSI_i = DSI - DSI_e.$$

Example 1.3

Determine the DKI and DSI of the structures shown in Figure 1.14a.

Solution

The nodes, elements, and active coordinates (i.e., direction of displacements and forces along the active DOF) of the structures are shown in Figure 1.14b. The sequence of numbering the DOF discussed in Section 1.6 is followed. In the figure, the reactions are indicated by arrows with slashed line.

Structure I

The axial deformations are usually neglected in the analysis of beams. Hence, there is only one active DOF (also see Table 1.1 to identify nodal DOF in different types of structures). Hence, DKI = 1.

Let N_n = number of nodes, N_e = number of elements, N_R = number of reaction components, and N_u = number of unknown forces (element forces + reactions)

$$N_n = 2, N_e = 1 \text{ and } N_R = 3$$

---

[*] In flexibility method, this statically determinate and stable structure is called the primary structure.

[†] A negative value of DSI indicates an unstable structure and a zero value of DSI represent a statically determinate structure.

[‡] Sometimes the elements of the structure are connected together by hinges or rollers etc. These connection conditions will provide additional equations for finding reactions. Such additional equations are known as equations of condition.

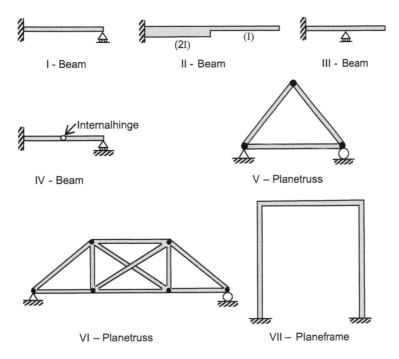

FIGURE 1.14a
Example 1.3.

In a beam element, the number of internal forces is two, namely bending moment and shear force. Hence, number of element forces $= 2N_e$ (also see Figure 1.3).

$$N_u = 2N_e + N_R = 2 + 3 = 5$$

Assuming that the structure lies the X–Y plane, using FBD of each node of the truss, two equations of equilibrium, namely $\Sigma F_Y = 0$ and $\Sigma M_Z = 0$, can be used to find the unknown forces. Hence, $N_E = 2N_n$.

$$N_E = 2 \times N_n = 2 \times 2 = 4$$

$$\text{DSI} = N_u - N_E = 5 - 4 = 1$$

Structure II

Since there is a change in the cross-section of the beam, the beam is modeled using two elements.

$$\text{DKI} = \text{number of active coordinates} = 4$$

$$N_n = 3, \, N_e = 2 \text{ and } N_R = 2$$

$$N_u = 2N_e + N_R = (2 \times 2) + 2 = 6$$

$$N_E = 2N_n = 2 \times 3 = 6$$

$$\text{DSI} = N_u - N_E = 6 - 6 = 0$$

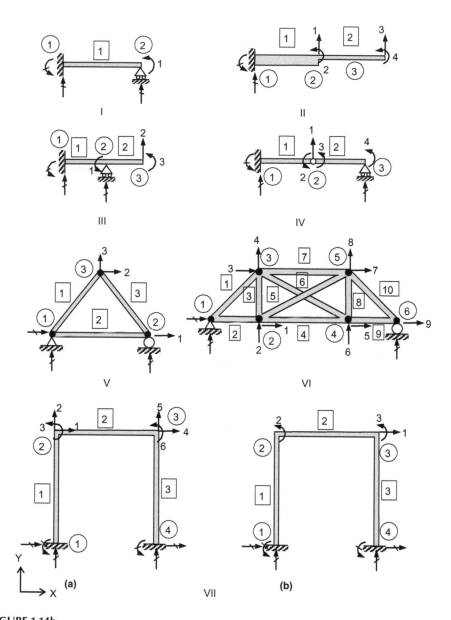

FIGURE 1.14b
Nodes, elements, active coordinates, and reactions. (a) Considering axial deformation. (b) Neglecting axial deformation.

Hence, the structure is statically determinate.

Structure III

$$DKI = 3$$

$$N_n = 3, N_e = 2, N_R = 3$$

$$N_u = 2N_n + N_R = (2 \times 2) + 3 = 7$$

$$N_E = 2N_n = 2 \times 3 = 6$$

$$\text{DSI} = N_u - N_E = 7 - 6 = 1$$

Structure IV

In this structure, though the vertical deflection at node 2 (at internal hinge) is the same for elements 1 and 2, the rotations will be different for both the elements. Hence, two active coordinates (2 and 3) are indicated to represent rotations at node 2.

$$\text{DKI} = 4$$

$$N_n = 3, \, N_e = 2, \, N_R = 3$$

$$N_u = 2N_e + N_R = (2 \times 2) + 3 = 7$$

Let N_c = Number of equations of condition
Here, $N_c = 1$ (bending moment at internal hinge is zero)
Hence, $N_E = 2N_n + N_c = (2 \times 3) + 1 = 7$

$$\text{DSI} = N_u - N_E = 7 - 7 = 0$$

The structure is statically determinate.

Structure V

$$\text{DKI} = 3$$

$$N_n = 3, \, N_e = 3, \, N_R = 3$$

In a truss element, there is only one internal force, namely axial force. Hence, the number of element forces = $1N_e$.

$$N_u = 1N_e + N_R = (1 \times 3) + 3 = 6$$

Using FBD of the node of truss, two equations of equilibrium, $\Sigma F_X = 0, \Sigma F_Y = 0$, are used to find the unknown forces. Hence, $N_E = 2N_n$.
   Therefore, $\text{DSI} = N_u - N_E = 6 - 6 = 0$

Structure VI

$$\text{DKI} = 9$$

$$N_n = 6, \, N_e = 10, \, N_R = 3$$

$$N_u = 1N_e + N_R = 10 + 3 = 13$$

$$N_E = 2N_n = 2 \times 6 = 12$$

$$\text{DSI} = 13 - 12 = 1$$

The equilibrium equations using the FBD of the structure are $\Sigma F_X = 0, \Sigma F_Y = 0$ and $\Sigma M_Z = 0$. Hence,

$$N_{ES} = \text{number of equilibrium equations for the structure} = 3$$

Therefore, $\text{DSI}_e = N_R - N_{ES} = 3 - 3 = 0$

Thus, the truss is statically determinate externally, i.e., the reactions can be determined using equations of equilibrium:

$$\text{DSI}_i = \text{DSI} - \text{DSI}_e = 1 - 0 = 1$$

The truss is statically indeterminate internally.

Structure VII

If axial deformations are considered, then DKI = 6.

When the axial deformations are neglected, the elements cannot elongate (or shorten). Since the ends of the elements 1 and 3 are fixed, there cannot be displacement along the active coordinates 2 and 6 shown in Figure 1.14b, VII(a). This is due to the fact that the length of the element does not change. Due to the same reason, the displacements along the active coordinate 1 will be the same as the displacement along the active coordinate 4. Hence, it is sufficient to consider only one active coordinate (either 1 or 4) to indicate sway of the frame. The active coordinates of the frame when the axial deformations are neglected are shown in Figure 1.14b, VII(b). In this case, DKI = 3.

$$N_n = 4, N_e = 3, N_R = 6$$

In a frame element, three internal forces exist, namely axial force, shear force, and bending moment. Hence, the number of element forces = $3N_e$

$$N_u = 3N_e + N_R = (3 \times 3) + 6 = 15$$

Using the FBD of each node of frame, three equations of equilibrium $\Sigma F_X = 0, \Sigma F_Y = 0$, and $\Sigma M_Z = 0$ are used to find unknown forces. Hence, $N_E = 3N_n$

$$N_E = 3N_n = 3 \times 4 = 12$$

Therefore, $\text{DSI} = N_u - N_E = 15 - 12 = 3$

1.11 Methods of Structural Analysis

Any method in structural analysis uses the three basic equations in mechanics: (i) *equilibrium equations*, (ii) *compatibility of displacements*, and (iii) *force–displacement relations*.

1.11.1 Equilibrium Equations

Consider the FBD of a body subjected to different forces. The resultant of any force system acting on a body can be expressed by a force vector $\mathbf{F_R}$ and couple $\mathbf{C_R}$ acting at a point. If the body is in equilibrium, then as per Newton's first law of motion, the resultant of the force system should be equal to zero, i.e., $\mathbf{F_R} = \mathbf{0}$ and $\mathbf{C_R} = \mathbf{0}$. These equations are known the *equations of equilibrium*.

In rectangular Cartesian coordinates, the equations of equilibrium can be written as follows:

$$\mathbf{F_R} = \mathbf{0} \Rightarrow \Sigma F_X = 0, \quad \Sigma F_Y = 0, \quad \Sigma F_Z = 0 \tag{1.4a}$$

$$\mathbf{C_R} = \mathbf{0} \Rightarrow \Sigma M_X = 0, \quad \Sigma M_Y = 0, \quad \Sigma M_Z = 0 \tag{1.4b}$$

where $\Sigma F_X = 0$, $\Sigma F_Y = 0$ and $\Sigma F_Z = 0$ are the algebraic sum of forces in the direction of X-, Y-, and Z-axes. Similarly, $\Sigma M_X = 0$, $\Sigma M_Y = 0$, and $\Sigma M_Z = 0$ represent the algebraic sum of moments about the X-, Y-, and Z-axes, respectively.

1.11.2 Compatibility of Displacements

These conditions refer to the continuity of displacement throughout the structure. The compatibility conditions are usually considered at the joints in a structure. For example, at a fixed support, there should be no translations and rotations, and at a rigid joint, all elements meeting at the joint should have same displacements (translations and rotations).

1.11.3 Force–Displacement Relations

In the case of a linearly elastic structure, the force–displacement relation is linear. Depending on the method of analysis, the force–displacement relation is expressed either in a flexibility or stiffness format. The force–displacement relations of structures and elements are discussed in detail in Chapter 2.

1.11.4 Flexibility and Stiffness Method of Analysis

All the methods of analysis of structures fall under two categories, namely the flexibility method (or force method) and the stiffness method (or displacement method).

In flexibility method, forces are taken as the basic unknowns. In this method, a statically indeterminate structure is made statically determinate (DSI = 0) by removing the excess or redundant forces (these forces are the basic unknowns and the number of unknown forces is equal to DSI). The statically determinate and stable structure thus obtained is known as the *primary structure*. The primary structure is then subjected to external forces (loads) and redundant forces, and compatibility conditions are used to find the redundant forces.

In stiffness method, displacements are chosen as the basic unknowns. A kinematically indeterminate structure is made kinematically determinate (DKI = 0)[*] by preventing all nodal displacements. These nodal displacements are in fact the basic unknowns and the number of unknown is equal to DKI. Then the unknown displacements are allowed, and equilibrium equations (using the FBD of nodes) are used to find the unknown displacements.

The choice of the method of analysis for a structure depends on the number of unknowns. If the DSI of the structure is less than DKI, then the flexibility method may be used and vice versa. However, it should be noted that the choice of choosing the redundant forces in the flexibility method is not unique. For example, some possible primary structures for a statically indeterminate structure (Figure 1.15) are shown in Figure 1.16.

In Figure 1.16a, the vertical reaction at B is chosen as the redundant force. The moment at support A is considered as redundant in Figure 1.16b and bending moment at C is the redundant force in Figure 1.16c. Thus, the solution using the flexibility method can be obtained using different ways by choosing different redundant forces. Hence, a systematic procedure cannot be developed for the analysis of structure using flexibility method.

---

[*] This kinematically determinate structure is the primary structure in stiffness method.

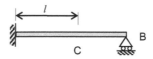

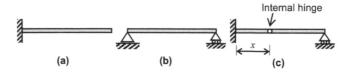

FIGURE 1.15
Statically indeterminate structure.

FIGURE 1.16
Possible primary structures.

On the other hand, in stiffness method, the choice of unknown displacements is unique for a structure. Hence, the whole procedure is systematic and is amenable to computer programming.

Problems

1.1 Determine the DSI and DKI of the framed structures shown in Figure 1.17.

1.2 Determine the nodal loads required for analyzing the framed structures shown in Figure 1.18.

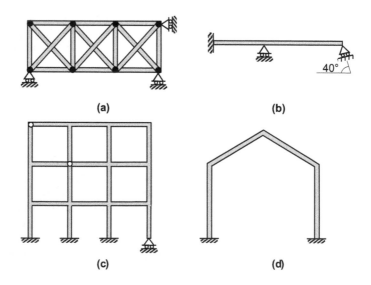

FIGURE 1.17
Problem 1.1.

27

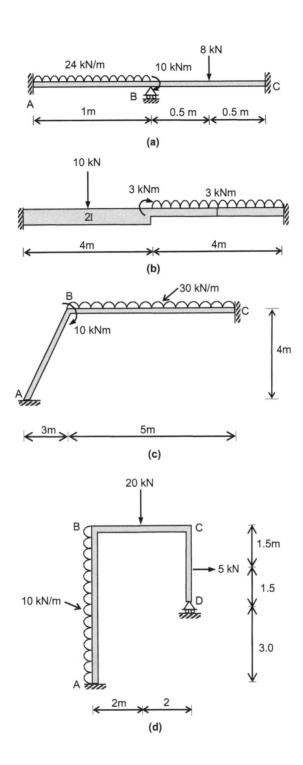

FIGURE 1.18
Problem 1.2.

2

Flexibility and Stiffness:
Characteristics of Structures

2.1 Introduction

The force–displacement relation of a structure is linear if it satisfies the following three conditions: (i) the material of the structure follows Hooke's law, i.e., the stress–strain relation is linear, (ii) the displacements of the structure are small, and (iii) there is no interaction between the flexural and axial affects in the constituent elements. Only structures having linear force–displacement relations are discussed in this chapter. The force–displacement relations are expressed using the flexibility and stiffness of the structure. In this chapter, the procedures to find the flexibility and stiffness of structures are developed. Also, the concept of work and energy, and their relations with the flexibility and stiffness are discussed.

2.2 Force–Displacement Relation of a Structure

The force–displacement relation is necessary for the analysis of the structure. The relation can be written either in terms of flexibility or stiffness coefficients. These coefficients are in fact the characteristics of the given structure and its coordinate system. The procedures for finding these coefficients for different types of structures are discussed in this section.

2.2.1 Structures with Single Coordinate

Figure 2.1 shows a structure with a single coordinate (the term coordinate indicates the direction of the displacement and the corresponding force at a point). In the figure, the linearly elastic spring is stretched by a force P and the displacement of the spring is D. The relation between P and D can be written using the *displacement equation* as

$$D = FP \qquad (2.1)$$

where F is the *flexibility* of the spring defined as the displacement caused by a unit value of the force P. From the equation, it can be seen that the flexibility F helps to transform the information in terms of force to the information in terms of displacements at the coordinate.

Alternatively, the force–displacement relation of the spring can be written using the *force equation* as

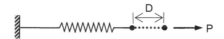

FIGURE 2.1
Linearly elastic spring.

$$P = KD \qquad (2.2)$$

where K is the *stiffness* of the spring defined as the force required to cause unit displacement. In the equation, K transforms the information in terms of displacement to information in terms of force.

Comparing Eqs. 2.1 and 2.2, it can be seen that the flexibility and stiffness of the spring are inverse to one another, i.e.,

$$F = \frac{1}{K} \text{ or } K = \frac{1}{F} \qquad (2.3a)$$

$$KF = 1 \qquad (2.3b)$$

Equations 2.1–2.3 for the linearly elastic spring are also valid for any linearly elastic structure with single coordinate. Figure 2.2 shows some structures having single coordinate. The flexibility and stiffness of the structure are also indicated in the figure.

2.2.2 Structures with Multiple Coordinates

Consider a structure with three coordinates shown in Figure 2.3. The forces (P_1, P_2, and P_3) and displacements (D_1, D_2, and D_3) are assumed to be positive, if they are in the same direction of the coordinates.

2.2.2.1 Flexibility Matrix

Since the force–displacement relationship is linear, the principle of superposition[*] can be used. Using this principle, it is observed that the response of the structure subjected to the loads P_1, P_2, and P_3 is equal to the sum of the response of the structure subjected to the loads P_1, P_2, and P_3 acting separately (Figure 2.4).

Thus, the displacements shown in Figure 2.3 can be written as the sum of displacements caused by individual loads (Figure 2.5). For example, the displacement along the first coordinate D_1 is given by

$$D_1 = D_{11} + D_{12} + D_{13} \qquad (a)$$

where D_{11}, D_{12}, and D_{13} are the displacements along the first coordinate due to loads acting at the coordinates 1, 2, and 3, respectively. In general, D_{ij} is the displacement at the ith coordinate due to loading at jth coordinate and the loads at the other coordinates are equal to zero. Similarly, the displacements D_2 and D_3 are

---

[*] The principle of superposition states that the effects produced by several causes can be obtained by combining the effects due to individual causes.

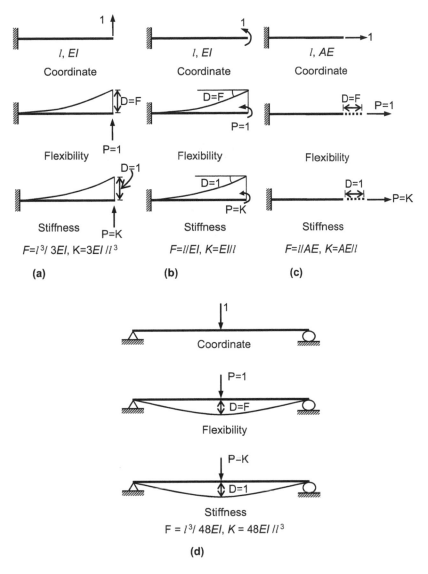

FIGURE 2.2
Flexibility and stiffness of structures with single coordinate: (a) structure 1, (b) structure 2, (c) structure 3, and (d) structure 4.

$$D_2 = D_{21} + D_{22} + D_{23} \tag{b}$$

$$D_3 = D_{31} + D_{32} + D_{33} \tag{c}$$

Figure 2.6a shows the deflection of the structure due to a unit load acting at the first coordinate. Comparing Figure 2.5a, it can be seen that $D_{11} = F_{11}P_1$, $D_{21} = F_{21}P_1$, and $D_{31} = F_{31}P_1$ (by using the principle of superposition). Similarly, the displacements in Figures 2.5c and d are written in terms of the displacements due to unit loads acting along the coordinates 2 and 3 (Figures 2.6b and c).

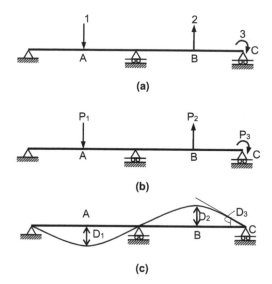

FIGURE 2.3
Structure with three coordinates: (a) coordinates, (b) forces, and (c) displacements.

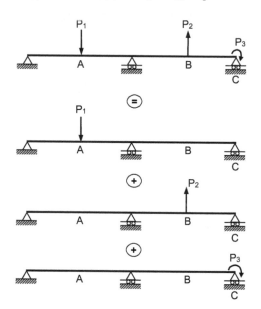

FIGURE 2.4
Application of superposition principle for loads.

In Figure 2.6, F_{ij} ($i, j = 1$ to 3) is the displacement at the ith coordinate due to a unit load at j^{th} coordinate and the loads at all other coordinates are equal to zero.

The Eqs. (a) to (c) in terms of F_{ij} are

$$D_1 = F_{11}P_1 + F_{12}P_2 + F_{13}P_3$$

$$D_2 = F_{21}P_1 + F_{22}P_2 + F_{23}P_3 \quad\quad\quad (d)$$

$$D_3 = F_{31}P_1 + F_{32}P_2 + F_{33}P_3$$

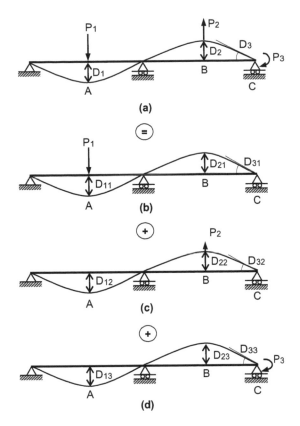

FIGURE 2.5
Illustration of superposition principle for displacements.

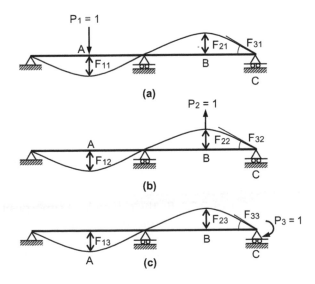

FIGURE 2.6
Flexibility coefficients.

Thus, for a structure having n coordinates, the above displacement equations can be written as

$$D_1 = F_{11}P_1 + F_{12}P_2 + F_{13}P_3 + \cdots + F_{1n}P_n$$

$$D_2 = F_{21}P_1 + F_{22}P_2 + F_{23}P_3 + \cdots + F_{2n}P_n$$

$$\cdots\cdots\cdots\cdots\cdots\cdots\cdots\cdots\cdots\cdots\cdots\cdots\cdots\cdots\cdots$$

(2.4a)

$$D_n = F_{n1}P_1 + F_{n2}P_2 + F_{n3}P_3 + \cdots + F_{nn}P_n$$

or

$$D_i = \sum_{j=1}^{n} F_{ij}P_j \tag{2.4b}$$

where P_1, P_2, ..., P_n and D_1, D_2, ..., D_n are the forces and displacements along the n coordinates.

In matrix notation, Eq. 2.4a is written as

$$\begin{Bmatrix} D_1 \\ D_2 \\ \cdots \\ D_n \end{Bmatrix} = \begin{bmatrix} F_{11} & F_{12} & \cdots & F_{1n} \\ F_{21} & F_{22} & \cdots & F_{2n} \\ \cdots & \cdots & \cdots & \cdots \\ F_{n1} & F_{n2} & \cdots & F_{nn} \end{bmatrix} \begin{Bmatrix} P_1 \\ P_2 \\ \cdots \\ P_n \end{Bmatrix}$$

or

$$\{D\} = [F]\{P\} \tag{2.5}$$

where $\{D\}$ is the *displacement vector* of order $n \times 1$, $[F]$ is a square matrix of order n known as the *flexibility matrix* and $\{P\}$ is the *force vector* of order $n \times 1$. The coefficients F_{11}, F_{12}, ..., in $[F]$ are known as the *flexibility coefficients*. The flexibility coefficients along the principal diagonal of $[F]$ are F_{11}, F_{22}, ... (F_{ij}, $i = j$) and they are the *direct flexibility coefficients*. The other flexibility coefficients (F_{ij}, $i \neq j$) are the *cross flexibility coefficients*.

2.2.2.2 Stiffness Matrix

In the previous case, the displacements were expressed in terms of forces using flexibility matrix. As an alternate way, similar to Eq. 2.2, the forces can be written in terms of displacements. To develop these equations, consider the response of the structure subjected to displacements D_1, D_2, and D_3 (Figure 2.3c). By the principle of superposition, this response is equal to the sum of response of structure subjected to displacements D_1, D_2, and D_3 acting separately (Figure 2.7).

In the figure, additional supports are introduced at the coordinates (A, B, and C) so as to get the appropriate deformed shape of the structure. The forces acting at the coordinates are written as the sum of forces caused by the displacements D_1, D_2, and D_3 applied one at a time (Figure 2.8). In the figure, P_{ij} (i, j = 1 to 3) is the force developed at the ith

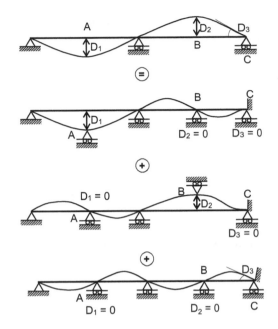

FIGURE 2.7
Application of superposition principle for displacements.

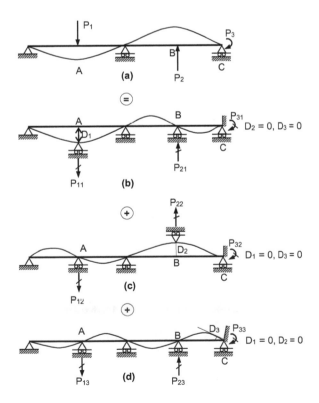

FIGURE 2.8
Illustration of superposition principle for forces.

coordinate due to the displacement at the jth coordinate, while the displacements at all other coordinates are equal to zero. In fact, P_{ij} are the reactions developed at the additional supports.

Thus, we have

$$P_1 = P_{11} + P_{12} + P_{13}$$

$$P_2 = P_{21} + P_{22} + P_{23} \qquad (e)$$

$$P_3 = P_{31} + P_{32} + P_{33}$$

The forces P_{ij} can be written in terms of forces (K_{ij}) caused by unit displacements at the coordinates (Figure 2.9).

In the figure, K_{ij} ($i, j = 1$ to 3) is the force developed at the ith coordinate due to unit displacement at the jth coordinate, while the displacements at other coordinates are zero. The forces K_{ij} can be also interpreted as the forces required to maintain the deformed configuration of the structure or as the reactions developed at the additional supports due to the application of unit displacement at a coordinate. From Figures 2.8b and 2.9a, it is seen that $P_{11} = K_{11}D_1$, $P_{21} = K_{21}D_1$, and $P_{31} = K_{31}D_1$. Similarly, the other forces in Figure 2.8 are written in terms of forces shown in Figure 2.9. Thus, Eq. e can be written as

$$P_1 = K_{11}D_1 + K_{12}D_2 + K_{13}D_3$$

$$P_2 = K_{21}D_1 + K_{22}D_2 + K_{23}D_3 \qquad (f)$$

$$P_3 = K_{31}D_1 + K_{32}D_2 + K_{33}D_3$$

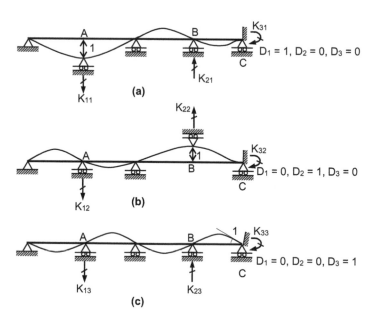

FIGURE 2.9
Stiffness coefficients.

For an n coordinate system, Eq. f can be written as

$$P_1 = K_{11}D_1 + K_{12}D_2 + K_{13}D_3 + \cdots + K_{1n}D_n$$

$$P_2 = K_{21}D_1 + K_{22}D_2 + K_{23}D_3 + \cdots + K_{2n}D_n$$

$$\cdots\cdots\cdots\cdots\cdots\cdots\cdots\cdots\cdots\cdots\cdots\cdots\cdots\cdots\cdots\cdots$$

$$P_n = K_{n1}D_1 + K_{n2}D_2 + K_{n3}D_3 + \cdots + K_{nn}D_n$$

(2.6a)

or

$$P_i = \sum_{j=1}^{n} K_{ij}D_j$$

(2.6b)

In matrix form,

$$\begin{Bmatrix} P_1 \\ P_2 \\ \cdots \\ P_n \end{Bmatrix} = \begin{bmatrix} K_{11} & K_{12} & \cdots & K_{1n} \\ K_{21} & K_{22} & \cdots & K_{2n} \\ \cdots & \cdots & \cdots & \cdots \\ K_{n1} & K_{n2} & \cdots & K_{nn} \end{bmatrix} \begin{Bmatrix} D_1 \\ D_2 \\ \cdots \\ D_n \end{Bmatrix}$$

or

$$\{P\} = [K]\{D\}$$

(2.7)

In the above equation, $[K]$ is a square matrix of order n and is the *stiffness matrix* of the structure. The coefficients K_{ij} are the *stiffness coefficients*. If $i = j$, then the stiffness coefficients are known as the *direct stiffness coefficients*, else they are called *cross stiffness coefficients*.

Comments

Similar to the flexibility method of analysis, in the stiffness method, it is necessary to have a primary structure for the analysis of structure. In the case of the flexibility method, the primary structure is a statically determinate and stable structure (i.e., DSI = 0). In the case of the stiffness method, the primary structure is a kinematically determinate structure

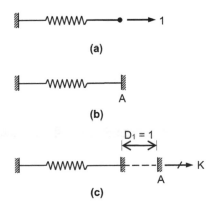

FIGURE 2.10
Stiffness of linear spring: (a) structure, (b) primary structure, and (c) reaction.

$$[F] = \begin{matrix} P_1=1 & P_2=1 & & P_i=1 & & P_j=1 & & P_n=1 \\ \begin{bmatrix} F_{11} & F_{12} & \cdots & F_{1i} & \cdots & F_{1j} & \cdots & F_{1n} \\ F_{21} & F_{22} & \cdots & F_{2i} & \cdots & F_{2j} & \cdots & F_{2n} \\ \cdots & \cdots & \cdots & \cdots & \cdots & \cdots & \cdots & \cdots \\ F_{i1} & F_{i2} & \cdots & F_{ii} & \cdots & F_{ij} & \cdots & F_{in} \\ \cdots & \cdots & \cdots & \cdots & \cdots & \cdots & \cdots & \cdots \\ F_{j1} & F_{j2} & \cdots & F_{ji} & \cdots & F_{jj} & \cdots & F_{jn} \\ \cdots & \cdots & \cdots & \cdots & \cdots & \cdots & \cdots & \cdots \\ F_{n1} & F_{n2} & \cdots & F_{ni} & \cdots & F_{nj} & \cdots & F_{nn} \end{bmatrix} \end{matrix}$$

FIGURE 2.12a
Flexibility matrix.

$$[K] = \begin{matrix} D_1=1 & D_2=1 & & D_i=1 & & D_j=1 & & D_n=1 \\ \begin{bmatrix} K_{11} & K_{12} & \cdots & K_{1i} & \cdots & K_{1j} & \cdots & K_{1n} \\ K_{21} & K_{22} & \cdots & K_{i2} & \cdots & K_{2j} & \cdots & K_{2n} \\ \cdots & \cdots & \cdots & \cdots & \cdots & \cdots & \cdots & \cdots \\ K_{i1} & K_{i2} & \cdots & K_{ii} & \cdots & K_{ij} & \cdots & K_{in} \\ \cdots & \cdots & \cdots & \cdots & \cdots & \cdots & \cdots & \cdots \\ K_{j1} & K_{j2} & \cdots & K_{ji} & \cdots & K_{jj} & \cdots & K_{jn} \\ \cdots & \cdots & \cdots & \cdots & \cdots & \cdots & \cdots & \cdots \\ K_{n1} & K_{n2} & \cdots & K_{ni} & \cdots & K_{nj} & \cdots & K_{nn} \end{bmatrix} \end{matrix}$$

FIGURE 2.12b
Stiffness matrix.

located at the ith row and j^{th} column of [K] is equal to the force developed at the ith coordinate due to unit displacement at the jth coordinate, while the displacements at all other coordinates are zero.

2.2.4 Properties of Flexibility and Stiffness Matrix

Some of the important properties of flexibility and stiffness matrices are listed below:

i. Both flexibility and stiffness matrices are square matrices of order n, where n is the number of coordinates.
ii. Flexibility matrix is the inverse of stiffness matrix.
From Eqs. 2.5 and 2.7,

$$\{D\} = [F]\{P\}$$

$$\Rightarrow \{P\} = [F]^{-1}\{D\} = [K]\{D\}$$

Hence,

$$[F]^{-1} = [K] \, or \, [K]^{-1}[F] \tag{2.8a}$$

$$\Rightarrow [F][K] = [I] \tag{2.8b}$$

where [I] is the identity matrix or unit matrix.

Thus, the flexibility matrix is the inverse of stiffness matrix. However, this does not indicate that the flexibility coefficient F_{ij} will be the inverse of stiffness coefficient K_{ij}.

iii. The elements on the principal diagonal of flexibility and stiffness matrices (F_{ii} and K_{ii}) are always positive, while the off-diagonal elements can be either positive or negative quantities.

F_{ii} is the displacement along the coordinate i due to a unit force at the ith coordinate. A force along a given direction, say along the positive direction of coordinate i, cannot cause displacement along the negative direction of coordinate i. Hence, F_{ii} is positive. K_{ii} is the force developed at the coordinate i due to unit displacement at the same coordinate. A force developed at coordinate i due to unit displacement at positive direction of coordinate i will be always in the positive direction of coordinate i. Thus, K_{ii} is positive.

iv. Flexibility and stiffness matrices are symmetric matrices, i.e.,

$$[F] = [F]^T \Rightarrow F_{ij} = F_{ji} \tag{2.9}$$

$$[K] = [K]^T \Rightarrow K_{ij} = K_{ji} \tag{2.10}$$

where $[F]^T$ and $[K]^T$ are the transpose of flexibility and stiffness matrices (proof is given in Section 2.4).

v. Flexibility matrix exists only for stable structures (see Example 2.2).

Example 2.1

Determine the flexibility and stiffness matrix for the structures shown in Figure 2.13a.

Solution

Structure 1

The structure has two coordinates. Hence, the flexibility and stiffness matrices are square matrices of order 2.

Flexibility matrix [F]

To find the deflections, the unit load method[*] is used. As per this method, the displacement D (translational or rotational) at a given point in a beam or frame is

$$D = \int \frac{Mmdx}{EI}$$

where M is the bending moment due to the applied loads and m is the bending moment due to the unit load (force or moment) applied at the point where it is necessary to find D.

For calculating displacement using the unit load method, the sagging bending moment is taken as positive (Figure 2.13b).

The flexibility coefficient F_{ij} is the displacement at i due to unit load applied at j (here applied load is also a unit load). Hence,

$$F_{ij} = \int \frac{M_j m_i dx}{EI}$$

---

[*] The reader may refer books on Basic Structural Analysis for getting more details of this method.

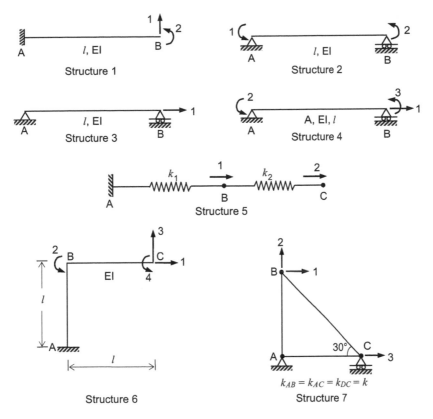

FIGURE 2.13a
Example 2.1.

FIGURE 2.13b
Sign convention for bending moments—unit load method.

where M_j is the bending moment due to load applied at j and m_i is the bending moment due to unit force applied at i. Since the load is having a unit value, $M_j = m_j$ (m_j is bending moment due to unit load at j). Thus,

$$F_{ij} = \int \frac{m_j m_i dx}{EI} = \int \frac{m_i m_j dx}{EI} = F_{ji}$$

The bending moments due to unit loads applied at the coordinates are given in Table 2.1 and Figure 2.13c.[*]

- Elements of first column of [F]
    For getting the elements, a unit load is applied at the first coordinate ($P_1 = 1$).
    Hence, $M_j = M_1$ ($M_1 = m_1$).

---

[*] In this book, the bending moment is drawn on the tension face of the element. This approach will be useful for the designer dealing with concrete structures.

TABLE 2.1

Bending Moment—Structure 1

Member	Origin	Limit	$m_1(x)$	$m_2(x)$
AB	B	0 to l	x	1

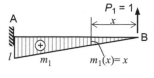 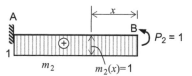

FIGURE 2.13c
Bending moment diagram (BMD)—Structure 1.

$$F_{11} = \int \frac{m_1 m_1 dx}{EI} = \int_0^l \frac{x^2 dx}{EI} = \frac{l^3}{3EI}$$

$$F_{21} = \int \frac{m_2 m_1 dx}{EI} = \int_0^l \frac{x dx}{EI} = \frac{l^2}{2EI}$$

- Elements of second column of [F]
  A unit load is applied at the second coordinate ($P_2 = 1$). In this case, $M_j = M_2 = m_2$.

$$F_{12} = F_{21} = l^2/2EI \quad \left(\text{due to symmetry of } [F]\right)$$

$$F_{22} = \int \frac{m_2 m_2 dx}{EI} = \int_0^l \frac{dx}{EI} = \frac{l}{EI}$$

The flexibility matrix of the structure is

$$[F] = \begin{bmatrix} l^2/3EI & l^2/2EI \\ l^2/2EI & l/EI \end{bmatrix}$$

Stiffness matrix [K]

To develop the stiffness matrix, it is necessary to start with the primary structure (DKI = 0). Since all the active coordinates are to be restrained, the primary structure required to develop [K] is the restrained structure. The structure and the restrained structure are shown in Figure 2.13d.

- Elements of first column of [K]
  To get the elements of first column of [K], give a unit displacement along the first coordinate ($D_1 = 1$) and make displacement along the second coordinate equal to zero ($D_2 = 0$). The forces developed in the member due to these displacements are obtained using the values given in Table 1.11. The stiffness coefficients are determined using the FBD of the joint B. Using equations of equilibrium,

$$K_{11} = 12EI/l^3 , \ K_{21} = -6EI/l^2$$

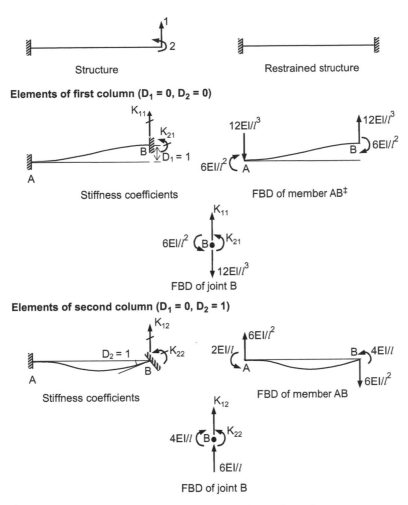

FIGURE 2.13d
Restrained structure and stiffness coefficients—Structure 1.

- Elements of second column of $[K]$
  To get the elements, give a unit displacement at the second coordinate (i.e., $D_2 = 1$ and $D_1 = 0$). Following the procedure described above, the stiffness coefficients are

$$K_{12} = -6EI/l^2 = K_{21}$$

$$K_{22} = 4EI/l$$

The stiffness matrix of the structure is

$$[K] = \begin{bmatrix} 12EI/l^3 & -6EI/l^2 \\ -6EI/l^2 & 4EI/l \end{bmatrix}$$

Structure 2
The flexibility and stiffness matrix are square matrices of order 2.

44 Matrix Methods of Structural Analysis

Flexibility Matrix

The bending moments due to unit loads at the coordinates are given in Table 2.2 and Figure 2.13e.

- Elements of first column ($P_1 = 1$, $P_2 = 0$)

$$F_{11} = \int \frac{m_1 m_1 dx}{EI} = \int_0^l \frac{\left(x/l - 1\right)^2 dx}{EI} = \frac{l}{3EI}$$

$$F_{21} = \int \frac{m_2 m_1 dx}{EI} = \int_0^l \frac{(x/l)(x/l - 1)dx}{EI} = \frac{-l}{6EI}$$

- Elements of second column ($P_1 = 0$, $P_2 = 1$)

$$F_{12} = \int \frac{m_1 m_2 dx}{EI} = F_{21} = \frac{-l}{6EI}$$

$$F_{22} = \int \frac{m_2 m_2 dx}{EI} = \int_0^l \frac{(x/l)^2 dx}{EI} = \frac{l}{3EI}$$

- Flexibility matrix [F]

$$[F] = \begin{bmatrix} l/3EI & -l/6EI \\ -l/6EI & l/3EI \end{bmatrix} = \frac{l}{6EI} \begin{bmatrix} 2 & -1 \\ -1 & 2 \end{bmatrix}$$

Stiffness Matrix

The restrained structure and the FBD of the member and joints are shown in Figure 2.13f.

- Elements of first column ($D_1 = 1$, $D_2 = 0$)
  From the FBD of joints A and B,

TABLE 2.2

Bending Moment—Structure 2

Member	Origin	Limit	$m_1(x)$	$m_2(x)$
AB	A	0 to l	$\dfrac{x}{l} - 1$	x/l

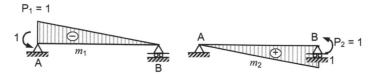

FIGURE 2.13e
BMD—Structure 2.

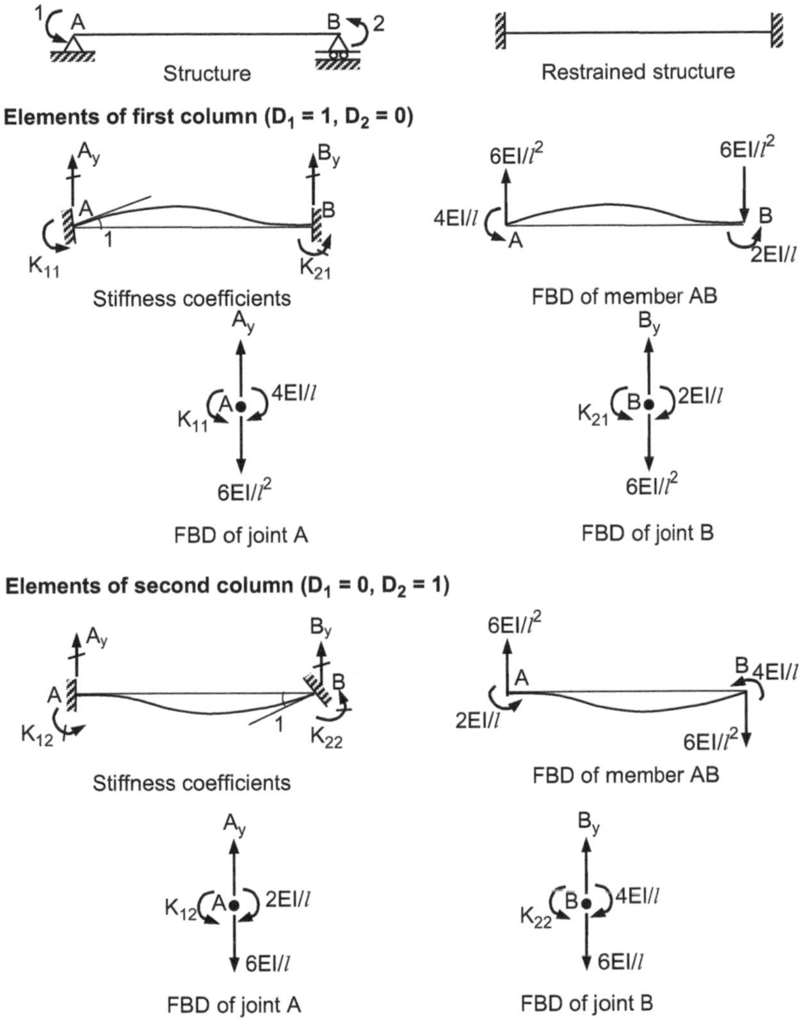

FIGURE 2.13f
Restrained structure and stiffness coefficients—Structure 2.

$$K_{11} = \frac{4EI}{l}, K_{21} = \frac{2EI}{l}$$

- Elements of second column ($D_1 = 0, D_2 = 1$)

  Using the FBD of joints A and B, $K_{12} = \frac{2EI}{l}, K_{22} = \frac{4EI}{l}$

- Stiffness matrix $[K]$

$$[K] = \begin{bmatrix} 4EI/l & 2EI/l \\ 2EI/l & 4EI/l \end{bmatrix} = 2EI/l \begin{bmatrix} 2 & 1 \\ 1 & 2 \end{bmatrix}$$

Structure 3

This is similar to Structure 3 shown in Figure 2.2. Hence, $K = AE/l$ and $F = l/AE$

Structure 4

Since there is no interaction between the axial and flexural deformations, the flexibility and stiffness matrices are developed using the superposition principle (Figure 2.13g). Hence,

$$[F] = \begin{bmatrix} l/AE & 0 & 0 \\ 0 & l/3EI & -l/6EI \\ 0 & -l/6EI & l/3EI \end{bmatrix} \text{ and } [K] = \begin{bmatrix} AE/l & 0 & 0 \\ 0 & 4EI/l & 2EI/l \\ 0 & 2EI/l & 4EI/l \end{bmatrix}$$

Structure 5

Flexibility Matrix

- Elements of first column ($P_1 = 1$, $P_2 = 0$)

    Let D_{AB} and D_{BC} are the elongations of the springs AB and BC, respectively. P_{AB} and P_{BC} are the forces in the two springs. From the FBD of the springs (Figure 2.13h),

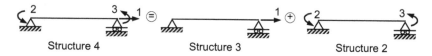

FIGURE 2.13g
Superposition principle—Structure 4.

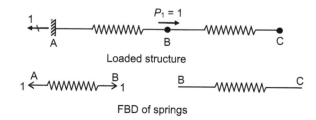

FIGURE 2.13h
FBD of springs.

$$P_{AB} = 1 \quad \Rightarrow D_{AB} = \frac{1}{k_1} P_{AB} = \frac{1}{k_1} \quad \left(\frac{1}{k_1} \text{ is the flexibility of the spring}\right)$$

$$P_{BC} = 0 \quad \Rightarrow D_{BC} = 0$$

$$\therefore \quad F_{11} = D_1 = D_{AB} = \frac{1}{k_1}$$

$$F_{21} = D_1 + D_2 = D_{AB} + D_{BC} = \frac{1}{k_1} + 0 = \frac{1}{k_1}$$

- Elements of second column ($P_1 = 0$, $P_2 = 1$)
  From Figure 2.13h,

$$P_{AB} = 1 \quad \Rightarrow D_{AB} = \frac{1}{k_1}$$

$$P_{BC} = 1 \quad \Rightarrow D_{BC} = \frac{1}{k_2}$$

Hence,

$$F_{12} = D_1 = D_{AB} = \frac{1}{k_1}$$

$$F_{22} = D_2 = D_{AB} + D_{BC} = \frac{1}{k_1} + \frac{1}{k_2}$$

- Flexibility matrix [F]

$$[F] = \begin{bmatrix} 1/k_1 & 1/k_1 \\ 1/k_1 & 1/k_1 + 1/k_2 \end{bmatrix}$$

Stiffness Matrix

- Elements of first column ($D_1 = 1$, $D_2 = 0$)
  From the FBD of nodes B and C (Figure 2.13i),

$$K_{11} = k_1 + k_2$$

$$K_{21} = -k_2$$

- Elements of second column ($D_1 = 0$, $D_2 = 1$)
  Using the FBD of nodes B and C (Figure 2.13i),

$$K_{12} = -k_2$$

$$K_{22} = k_2$$

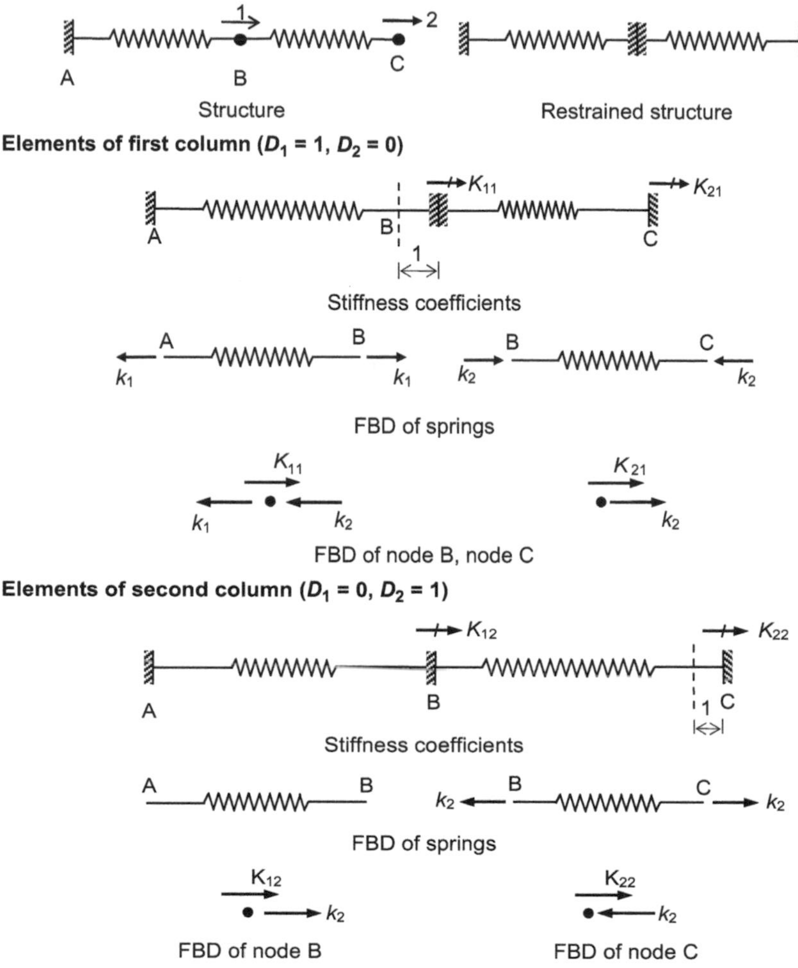

Elements of first column ($D_1 = 1, D_2 = 0$)

Stiffness coefficients

FBD of springs

FBD of node B, node C

Elements of second column ($D_1 = 0, D_2 = 1$)

Stiffness coefficients

FBD of springs

FBD of node B   **FBD of node C**

FIGURE 2.13i
Restrained structure and stiffness coefficients—Structure 5.

- Stiffness matrix $[K]$

$$[K] = \begin{bmatrix} k_1 + k_2 & -k_2 \\ -k_2 & k_2 \end{bmatrix}$$

Structure 6

The order of $[K]$ and $[F]$ is 4.

Flexibility Matrix

The bending moments due to unit loads applied at the coordinates are shown in Figure 2.13j. The bending moments in the different members are given in Table 2.3.

- Elements of first column ($P_1 = 1, P_2 = P_3 = P_4 = 0$)

$$F_{11} = \int \frac{m_1 m_1 dx}{EI} = \int_{BC} \frac{m_1 m_1 dx}{EI} + \int_{AB} \frac{m_1 m_1 dx}{EI} = 0 + \int_0^l \frac{(-x)^2 dx}{EI} = \frac{l^3}{3EI}$$

TABLE 2.3

Bending Moment—Structure 6

Member	Origin	Limit	m_1	m_2	m_3	m_4
BC	C	0 to l	0	0	x	1
AB	B	0 to l	$-x$	1	l	1

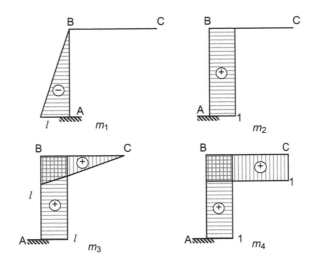

FIGURE 2.13j
BMD—Structure 6.

$$F_{21} = \int \frac{m_2 m_1 dx}{EI} = \int_0^l \frac{(-x)dx}{EI} = \frac{-l^2}{2EI}$$

$$F_{31} = \int \frac{m_3 m_1 dx}{EI} = \int_0^l \frac{(-lx)dx}{EI} = \frac{-l^3}{2EI}$$

$$F_{41} = \int \frac{m_4 m_1 dx}{EI} = \int_0^l \frac{(-x)dx}{EI} = \frac{-l^2}{2EI}$$

- Elements of second column ($P_2 = 1$, $P_1 = P_3 = P_4 = 0$)

$$F_{12} = F_{21} = \frac{-l^2}{2EI}$$

$$F_{22} = \int \frac{m_2 m_2 dx}{EI} = \int_0^l \frac{dx}{EI} = \frac{l}{EI}$$

$$F_{32} = \int \frac{m_3 m_2 dx}{EI} = \int_0^l \frac{ldx}{EI} = \frac{l^2}{EI}$$

$$F_{42} = \int \frac{m_4 m_2 dx}{EI} = \int_0^l \frac{dx}{EI} = \frac{l}{EI}$$

- Elements of third column ($P_3 = 1$, $P_1 = P_2 = P_4 = 0$)

$$F_{13} = F_{31} = \frac{-l^3}{2EI}$$

$$F_{23} = F_{32} = \frac{l^2}{EI}$$

$$F_{33} = \int \frac{m_3 m_3 dx}{EI} = \int_0^l \frac{x^2 dx}{EI} + \int_0^l \frac{l^2 dx}{EI} = \frac{4l^3}{3EI}$$

$$F_{43} = \int \frac{m_4 m_3 dx}{EI} = \int_0^l \frac{x dx}{EI} + \int_0^l \frac{l dx}{EI} = \frac{3l^2}{2EI}$$

- Elements of fourth column ($P_4 = 1$, $P_1 = P_2 = P_3 = 0$)

$$F_{14} = F_{41} = \frac{-l^2}{2EI}$$

$$F_{24} = F_{42} = \frac{l}{EI}$$

$$F_{34} = F_{43} = \frac{3l^2}{2EI}$$

$$F_{44} = \int \frac{m_4 m_4 dx}{EI} = \int_0^l \frac{dx}{EI} + \int_0^l \frac{dx}{EI} = \frac{2l}{EI}$$

- Flexibility matrix

$$[F] = \begin{bmatrix} l^3/3EI & -l^2/2EI & -l^3/2EI & -l^2/2EI \\ -l^2/2EI & l/EI & l^2/EI & l/EI \\ -l^3/2EI & l^2/EI & 4l^3/3EI & 3l^2/2EI \\ -l^2/2EI & l/EI & 3l^2/2EI & 2l/EI \end{bmatrix}$$

Stiffness Matrix

- Elements of first column ($D_1 = 1$, $D_2 = D_3$, and $D_4 = 0$) (Figure 2.13k)
  From FBD of joint B,

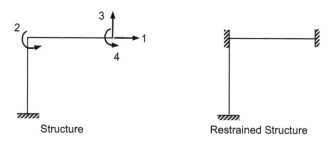

Structure   Restrained Structure

FIGURE 2.13k
Restrained structure and stiffness coefficients—Structure 6.

(Continued)

Elements of first column ($D_1 = 1, D_2 = D_3 = D_4 = 0$)

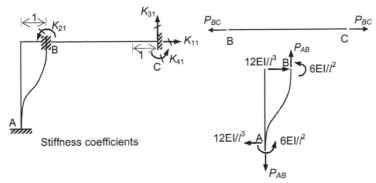

Stiffness coefficients

FBD of members AB and BC

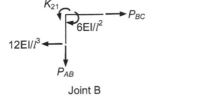

Joint B

Joint C

FBD of joints

Elements of second column ($D_2 = 1, D_1 = D_3 = D_4 = 0$)

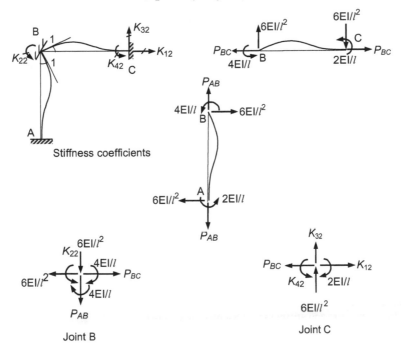

Stiffness coefficients

Joint B

Joint C

FBD of joints B and C

FIGURE 2.13k (CONTINUED)
Restrained structure and stiffness coefficients—Structure 6.

(Continued)

Elements of third column ($D_3 = 1$, $D_1 = D_2 = D_4 = 0$)

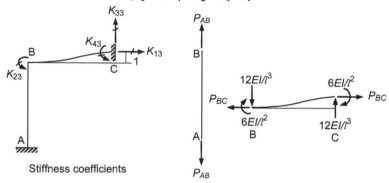

Stiffness coefficients   FBD of members AB and BC

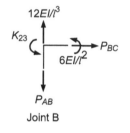

Joint B

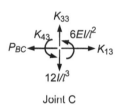

Joint C

Elements of fourth column ($D_4 = 1$, $D_1 = D_2 = D_3 = 0$)

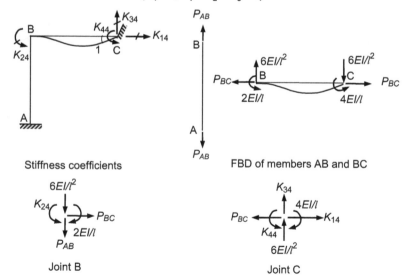

Stiffness coefficients   FBD of members AB and BC

Joint B   Joint C

FIGURE 2.13k (CONTINUED)
Restrained structure and stiffness coefficients—Structure 6.

$$P_{BC} = 12EI/l^3$$

$$K_{21} = 6EI/l^2$$

From FBD of joint C,

$$K_{31} = 0$$

$$K_{41} = 0$$

$$K_{11} = P_{BC} = 12EI/l^3$$

Note: Since the given structure is a plane frame, axial forces (P_{AB}, P_{BC}) are developed in the structure.
- Elements of second column ($D_2 = 1$, $D_1 = D_3 = D_4 = 0$)
  From FBD of joint B,

$$P_{BC} = 6EI/l^2$$

$$K_{22} = 8EI/l$$

From FBD of joint C,

$$K_{32} = -6EI/l^2$$

$$K_{42} = 2EI/l$$

$$K_{12} = P_{BC} = 6EI/l^2$$

- Elements of third column ($D_3 = 1$, $D_1 = D_2 = D_4 = 0$)
  From FBD of joint B,

$$P_{BC} = 0$$

$$K_{23} = -6EI/l^2$$

From FBD of joint C,

$$K_{33} = 12EI/l^3$$

$$K_{43} = -6EI/l^2$$

$$K_{13} = P_{BC} = 0$$

- Elements of fourth column ($D_4 = 1$, $D_1 = D_2 = D_3 = 0$)
  From FBD of joint B,

$$P_{BC} = 0$$

$$K_{24} = 2EI/l$$

From FBD of joint C,

$$K_{34} = -6EI/l^2$$

$$K_{44} = 4EI/l$$

$$K_{14} = P_{BC} = 0$$

- Stiffness matrix

$$[K] = \begin{bmatrix} 12EI/l^3 & 6EI/l^2 & 0 & 0 \\ 6EI/l^2 & 8EI/l & -6EI/l^2 & 2EI/l \\ 0 & -6EI/l^2 & 12EI/l^3 & -6EI/l^2 \\ 0 & 2EI/l & -6EI/l^2 & 4EI/l \end{bmatrix}$$

Structure 7

The flexibility and stiffness matrices are square matrices of order 3.

Flexibility Matrix

The displacement at any point in the truss is determined using the equation (the unit load method),

$$D = \sum_n \frac{Ppl}{AE}$$

where P is the force in the truss member due to applied loads, p is the force in the truss member due to unit load applied at the point where D is to be found out, l is the length of truss member, A is the cross-sectional area of member, and n is the number of truss members.

Hence, the flexibility coefficient F_{ij} is

$$F_{ij} = \sum \frac{P_j p_i l}{AE}$$

Since the applied load is of unit value, $P_j = p_j$. Thus, we have

$$F_{ij} = \sum \frac{p_j p_i l}{AE} = \sum \frac{p_i p_j l}{AE} = F_{ji}$$

where p_i and p_j are the forces in the truss members due to unit loads applied at i and j, respectively.

The forces in the truss members due to unit loads applied in different coordinates are shown in Figure 2.13l and are also tabulated in Table 2.4. In the table, the negative value indicates that the forces are compressive in nature, else they are tensile in nature.

- Elements of first column

$$F_{11} = \sum \frac{p_1 p_1 l}{AE}$$

TABLE 2.4

Forces in the Members of Truss

Member	p_1	p_2	p_3
AB	$1/\sqrt{3}$	1	0
BC	$-2/\sqrt{3}$	0	0
AC	1	0	1

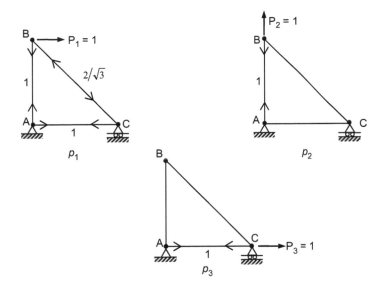

FIGURE 2.13l
Forces in truss—Structure 7.

For all members, $AE/l = k$

$$\Rightarrow \quad F_{11} = \frac{1}{k}\left[\frac{1}{3} + \frac{4}{3} + 1\right] = \frac{8}{3}\frac{1}{k}$$

$$F_{21} = \sum \frac{p_2 p_1 l}{AE} = \frac{1}{k}\left[\frac{1}{\sqrt{3}} + 0 + 0\right] = \frac{1}{\sqrt{3}}\frac{1}{k}$$

$$F_{31} = \sum \frac{p_3 p_1 l}{AE} = \frac{1}{k}[0 + 0 + 1] = \frac{1}{k}$$

- Elements of second column

$$F_{12} = F_{21} = \frac{1}{\sqrt{3}}\frac{1}{k}$$

$$F_{22} = \sum \frac{p_2 p_2 l}{AE} = \frac{1}{k}[1 + 0 + 0] = \frac{1}{k}$$

$$F_{32} = \sum \frac{p_3 p_2 l}{AE} = \frac{1}{k}[0 + 0 + 0] = 0$$

- Elements of third column

$$F_{13} = F_{31} = \frac{1}{k}$$

$$F_{23} = F_{32} = 0$$

$$F_{33} = \sum \frac{p_3 p_3 l}{AE} = \frac{1}{k}[0 + 0 + 1] = \frac{1}{k}$$

- Flexibility matrix

$$[F] = \frac{1}{k} \begin{bmatrix} 8/3 & 1/\sqrt{3} & 1 \\ 1/\sqrt{3} & 1 & 0 \\ 1 & 0 & 1 \end{bmatrix}$$

Stiffness Matrix

The restrained structure, stiffness coefficients, deformed shape, FBD of members, and joints are shown in Figure 2.13m.

- Elements of first column ($D_1 = 1$, $D_2 = D_3 = 0$)

  Let D_{AB}, D_{BC}, and D_{AC} are the elongations of the members AB, BC, and AC, respectively. A negative value indicates shortening of the members. The force developed in the members due to elongation is equal to the product of stiffness of the members (k) and the elongation of the member. The elongations and forces developed in the members are shown in the figure.

  From FBD of joint B,

$$K_{11} = k \frac{\sqrt{3}}{2} \cos 30 = \frac{3}{4} k$$

$$K_{21} = -k \frac{\sqrt{3}}{2} \sin 30 = -\frac{\sqrt{3}}{4} k$$

From FBD of joint C,

$$K_{31} = -k \frac{\sqrt{3}}{2} \cos 30 = -\frac{3}{4} k$$

- Elements of second column ($D_2 = 1$, $D_1 = D_3 = 0$)

  From FBD of joint B,

$$K_{12} = -\frac{k}{2} \cos 30 = -\frac{\sqrt{3}}{4} k$$

$$K_{22} = k + \frac{k}{2} \sin 30 = \frac{5}{4} k$$

From FBD of joint C,

$$K_{32} = \frac{k}{2} \cos 30 = \frac{\sqrt{3}}{4} k$$

- Elements of third column ($D_3 = 1$, $D_1 = D_2 = 0$)

  From FBD of joint B,

$$K_{13} = -\frac{\sqrt{3}}{2} k \cos 30 = -\frac{3}{4} k$$

$$K_{23} = \frac{\sqrt{3}}{2} k \sin 30 = \frac{\sqrt{3}}{4} k$$

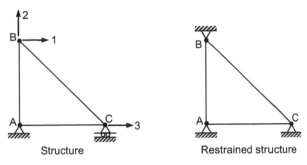

Elements of first column ($D_1 = 1$, $D_2 = D_3 = 0$)

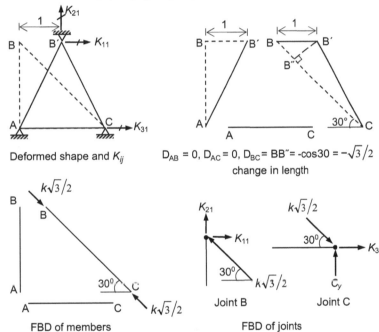

$D_{AB} = 0$, $D_{AC} = 0$, $D_{BC} = BB'' = -\cos30 = -\sqrt{3}/2$
change in length

Elements of second column ($D_2 = 1$, $D_1 = D_3 = 0$)

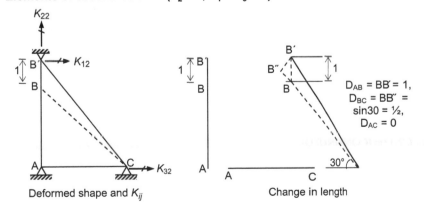

$D_{AB} = BB' = 1$,
$D_{BC} = BB' = \sin30 = \frac{1}{2}$,
$D_{AC} = 0$

FIGURE 2.13m
Restrained structure and stiffness coefficients—Structure 7.

(*Continued*)

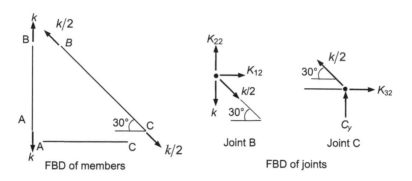

Elements of third column ($D_3 = 1$, $D_1 = D_2 = 0$)

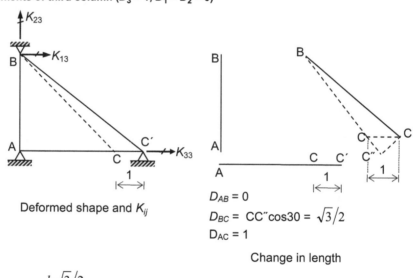

$D_{AB} = 0$

$D_{BC} = CC''\cos30 = \sqrt{3}/2$

$D_{AC} = 1$

Change in length

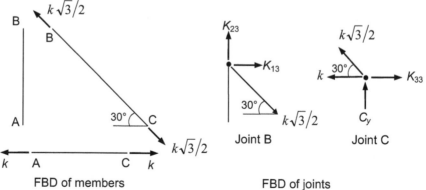

FIGURE 2.13m (CONTINUED)
Restrained structure and stiffness coefficients—Structure 7.

From FBD of joint C,

$$K_{33} = k + \frac{\sqrt{3}}{2}k\cos 30 = \frac{7}{4}k$$

- Stiffness matrix

$$[K] = k \begin{bmatrix} 3/4 & -\sqrt{3}/4 & -3/4 \\ -\sqrt{3}/4 & 5/4 & \sqrt{3}/4 \\ -3/4 & \sqrt{3}/4 & 7/4 \end{bmatrix}$$

Comments:

- The reader may verify that $[K][F] = [I]$ for all the structures.
- From the above examples, it can be seen that development of $[K]$ and $[F]$ for the structures using the classical approach, i.e., considering the behavior of the whole structure (see Structures 6 and 7) is difficult and time-consuming. However, element approach can be conveniently used for the development of flexibility and stiffness matrices of the structures and the procedures for the same are discussed in the later chapters.

Example 2.2

Develop the stiffness matrix for the following elements (Figure 2.14a):

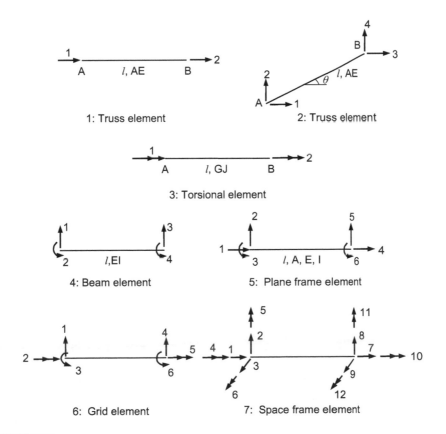

FIGURE 2.14a
Example 2.2.

Solution

Element 1

The restrained structure, FBD of element, and nodes are shown in Figure 2.14b.

- Elements of first column ($D_1 = 1$, $D_2 = 0$)
  Change in length of member $D_{AB} = -1$ (shortening of element)
  ∴ Force developed in the element $P_{AB} = -AE/l$ (compressive force)
  From FBD of node A,

$$K_{11} = AE/l$$

From FBD of node B,

$$K_{21} = -AE/l$$

- Elements of second column of ($D_2 = 1$, $D_1 = 0$)
  $D_{AB} = 1$ (elongation of member)
  $P_{AB} = AE/l$ (tensile force)
  From FBD of node A,

$$K_{12} = -AE/l$$

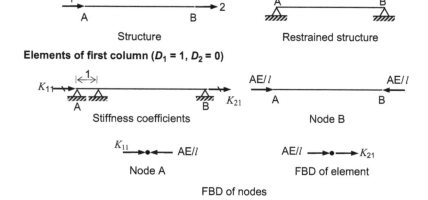

Elements of first column ($D_1 = 1$, $D_2 = 0$)

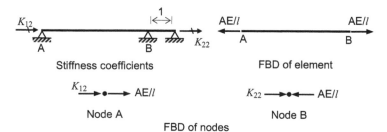

Elements of second column ($D_1 = 0$, $D_2 = 1$)

FIGURE 2.14b
Stiffness coefficients—Element 1.

From FBD of node B,

$$K_{22} = AE/l$$

- Stiffness matrix

$$[K] = \begin{bmatrix} AE/l & -AE/l \\ -AE/l & AE/l \end{bmatrix} = \frac{AE}{l} \begin{bmatrix} 1 & -1 \\ -1 & 1 \end{bmatrix}$$

Element 2

- Elements of first column ($D_1 = 1$, $D_2 = D_3 = D_4 = 0$) (Figure 2.14c)

$$D_{AB} = A'B - AB = A''B - AB = -AA'' = -\cos\theta = -c$$

$$P_{AB} = \frac{AE}{l} D_{AB} = -\frac{AE}{l} c$$

From FBD of node A,

$$K_{11} = \frac{AE}{l} c \times \cos\theta = \frac{AE}{l} c^2$$

$$K_{21} = \frac{AE}{l} c \times \sin\theta = \frac{AE}{l} cs \quad (\text{where } s = \sin\theta)$$

From FBD of node B,

$$K_{31} = -\frac{AE}{l} c \times \cos\theta = -\frac{AE}{l} c^2$$

$$K_{41} = -\frac{AE}{l} c \times \sin\theta = -\frac{AE}{l} cs$$

- Elements of second column ($D_2 = 1$, $D_1 = D_3 = D_4 = 0$)

$$D_{AB} = A'B - AB = A''B - AB = -AA'' = -s$$

$$P_{AB} = -\frac{AE}{l} s$$

From FBD of node A,

$$K_{12} = \frac{AE}{l} s \times \cos\theta = \frac{AE}{l} cs$$

$$K_{22} = \frac{AE}{l} s \times \sin\theta = \frac{AE}{l} s^2$$

From FBD of node B,

$$K_{32} = -\frac{AE}{l} s \times \cos\theta = -\frac{AE}{l} cs$$

$$K_{42} = -\frac{AE}{l} s \times \sin\theta = -\frac{AE}{l} s^2$$

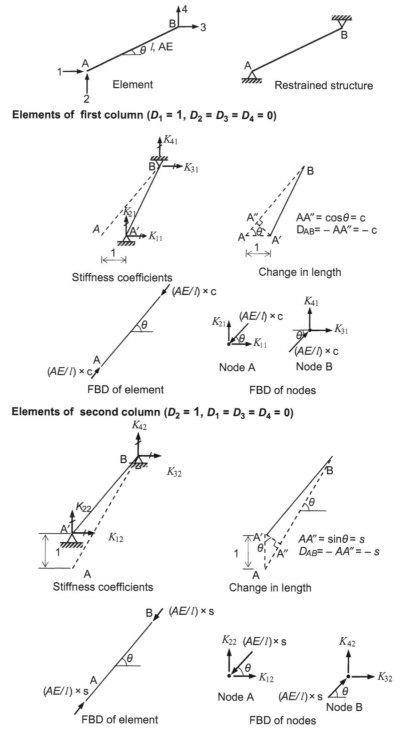

FIGURE 2.14c
Stiffness coefficients—Element 2.

(*Continued*)

Elements of third column ($D_3 = 1$, $D_1 = D_2 = D_4 = 0$)

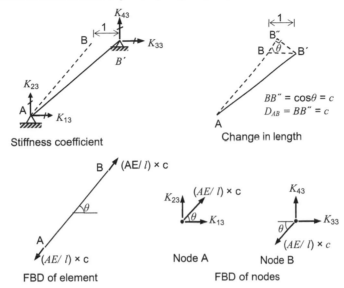

Stiffness coefficient   Change in length

$$BB'' = \cos\theta = c$$
$$D_{AB} = BB'' = c$$

FBD of element   Node A   Node B   FBD of nodes

Elements of fourth column ($D_4 = 1$, $D_1 = D_2 = D_3 = 0$)

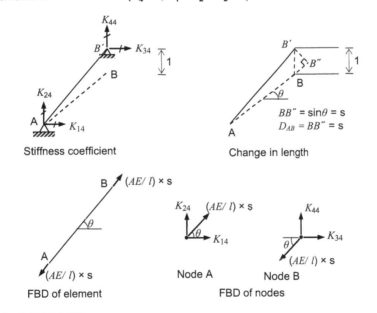

Stiffness coefficient   Change in length

$$BB'' = \sin\theta = s$$
$$D_{AB} = BB'' = s$$

FBD of element   Node A   Node B   FBD of nodes

FIGURE 2.14c (CONTINUED)
Stiffness coefficients—Element 2.

- Elements of third column ($D_3 = 1$, $D_1 = D_2 = D_4 = 0$)

$$D_{AB} = AB' - AB = AB'' - AB = BB'' = c$$

$$P_{AB} = \frac{AE}{l}c$$

From FBD of node A,

$$K_{13} = -\frac{AE}{l} c \times \cos\theta = -\frac{AE}{l} c^2$$

$$K_{23} = -\frac{AE}{l} c \times \sin\theta = -\frac{AE}{l} cs$$

From FBD of node B,

$$K_{33} = \frac{AE}{l} c \times \cos\theta = \frac{AE}{l} c^2$$

$$K_{43} = \frac{AE}{l} c \times \sin\theta = \frac{AE}{l} cs$$

- Elements of fourth column ($D_4 = 1$, $D_1 = D_2 = D_3 = 0$)

$$D_{AB} = AB' - AB = AB'' - AB = BB'' = s$$

$$P_{AB} = \frac{AE}{l} s$$

From FBD of node A,

$$K_{14} = -\frac{AE}{l} s \times \cos\theta = -\frac{AE}{l} cs$$

$$K_{24} = -\frac{AE}{l} s \times \sin\theta = -\frac{AE}{l} s^2$$

From FBD of node B,

$$K_{34} = \frac{AE}{l} s \times \cos\theta = \frac{AE}{l} cs$$

$$K_{44} = \frac{AE}{l} s \times \sin\theta = \frac{AE}{l} s^2$$

- Stiffness matrix

$$[K] = \frac{AE}{l} \begin{bmatrix} c^2 & cs & -c^2 & -cs \\ cs & s^2 & -cs & -s^2 \\ -c^2 & -cs & c^2 & cs \\ -cs & -s^2 & cs & s^2 \end{bmatrix}$$

Element 3

Element 3 is a torsional element which is used to model structures subjected to pure torsion. In Figure 2.14d, G is the shear modulus and J is the torsion constant. J for typical cross-sections is given in Table 2.5.

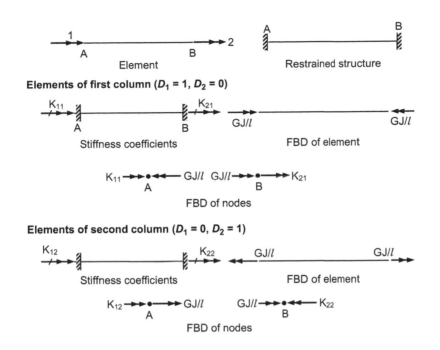

FIGURE 2.14d
Stiffness coefficients of Element 3.

TABLE 2.5

Torsion Constant

Section		Torsion Constant J
Solid circular section		$\dfrac{1}{2}\pi a^4$
Hollow circular section		$\dfrac{1}{2}\pi\left(a_i^4 - a_0^4\right)$
Thin-walled closed section		$4A_0^2 t/\Gamma$
Rectangular (b = smaller dimension)		$\dfrac{ab^3}{3}\left[1 - 0.63\dfrac{b}{a}\left(1 - \dfrac{b^4}{12a^4}\right)\right]$

A_0 = area enclosed by the median line.
Γ = length of the median line.

- Elements of first column ($D_1 = 1$, $D_2 = 0$)
  From FBD of nodes A and B,

$$K_{11} = GJ/l$$

$$K_{21} = GJ/l$$

- Elements of first column ($D_1 = 0$, $D_2 = 1$)
  From FBD of nodes,

$$K_{12} = -GJ/l$$

$$K_{22} = GJ/l$$

- Stiffness matrix $[K]$

$$[K] = \begin{bmatrix} GJ/l & -GJ/l \\ -GJ/l & GJ/l \end{bmatrix} = \frac{GJ}{l} \begin{bmatrix} 1 & -1 \\ -1 & 1 \end{bmatrix}$$

Element 4

- Elements of first column ($D_1 = 1$, $D_2 = D_3 = D_4 = 0$)
  From FBD of node A (Figure 2.14e),

$$K_{11} = 12EI/l^3$$

$$K_{21} = 6EI/l^2$$

From FBD of node B,

$$K_{31} = -12EI/l^3$$

$$K_{41} = 6EI/l^2$$

- Elements of second column ($D_2 = 1$, $D_1 = D_3 = D_4 = 0$)
  Using FBD of nodes,

$$K_{12} = 6EI/l^2 , \; K_{22} = 4EI/l, \; K_{32} = -6EI/l^2 , \; K_{42} = 2EI/l$$

- Elements of third column ($D_3 = 1$, $D_1 = D_2 = D_4 = 0$)
  From FBD of nodes,

$$K_{13} = -12EI/l^3 , \; K_{23} = -6EI/l^2 , \; K_{33} = 12EI/l^3 , \; K_{43} = -6EI/l^2$$

- Elements of fourth column ($D_4 = 1$, $D_1 = D_2 = D_3 = 0$)
  From FBD of nodes,

$$K_{14} = 6EI/l^2 , \; K_{24} = 2EI/l, \; K_{34} = -6EI/l^2 , \; K_{44} = 4EI/l$$

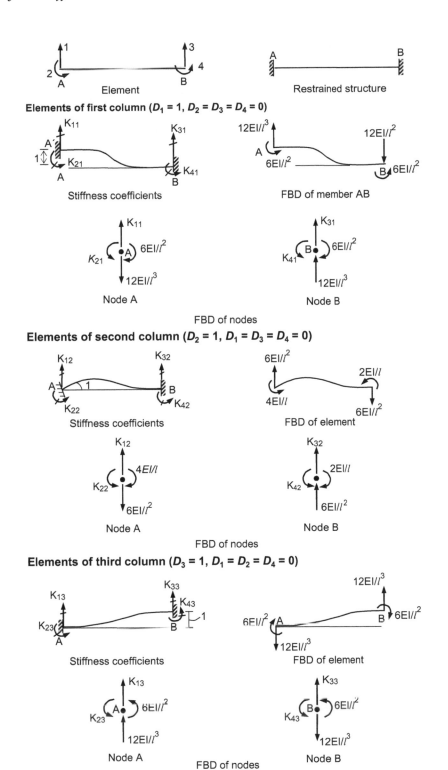

FIGURE 2.14e
Stiffness coefficients—Element 4.

(Continued)

Elements of fourth column ($D_4 = 1$, $D_1 = D_2 = D_3 = 0$)

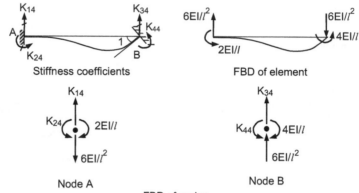

Stiffness coefficients   FBD of element

Node A   Node B

FBD of nodes

FIGURE 2.14e (CONTINUED)
Stiffness coefficients—Element 4.

- Stiffness matrix

$$[K] = \begin{bmatrix} 12EI/l^3 & 6EI/l^2 & -12EI/l^3 & 6EI/l^2 \\ 6EI/l^2 & 4EI/l & -6EI/l^2 & 2EI/l \\ -12\,EI/l^3 & -6EI/l^2 & 12EI/l^3 & -6EI/l^2 \\ 6EI/l^2 & 2EI/l & -6EI/l^2 & 4EI/l \end{bmatrix}$$

Element 5

There is no interaction between the flexural and axial effects in the element. Thus, [K] for the plane frame element is developed using the superposition principle (Figure 2.14f).
    Hence, we have

$$[K] = \begin{bmatrix} AE/l & 0 & 0 & -AE/l & 0 & 0 \\ 0 & 12EI/l^3 & 6EI/l^2 & 0 & -12\,EI/l^3 & 6EI/l^2 \\ 0 & 6EI/l^2 & 4EI/l & 0 & -6EI/l^2 & 2EI/l \\ -AE/l & 0 & 0 & AE/l & 0 & 0 \\ 0 & -12EI/l^3 & -6EI/l^2 & 0 & 12EI/l^3 & -6EI/l^2 \\ 0 & 6EI/l^2 & 2EI/l & 0 & -6EI/l^2 & 4EI/l \end{bmatrix}$$

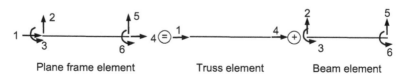

Plane frame element   Truss element   Beam element

FIGURE 2.14f
Superposition principle for plane frame element.

Element 6

The torsional deformations are independent of flexural deformations. Hence, [K] for the grid element is obtained using the superposition principle shown in Figure 2.14g.

Thus, we obtain

$$[K] = \begin{bmatrix} 12EI/l^3 & 0 & 6EI/l^2 & -12EI/l^3 & 0 & 6EI/l^2 \\ 0 & GJ/l & 0 & 0 & -GJ/l & 0 \\ 6EI/l^2 & 0 & 4EI/l & -6EI/l^2 & 0 & 2EI/l \\ -12EI/l^3 & 0 & -6EI/l^2 & 12EI/l^3 & 0 & -6EI/l^2 \\ 0 & -GJ/l & 0 & 0 & GJ/l & 0 \\ 6EI/l^2 & 0 & 2EI/l & -6EI/l^2 & 0 & 4EI/l \end{bmatrix}$$

Element 7

Let P_i and D_i ($i = 1$ to 12) are the element force and the corresponding displacement along the ith coordinate (Figures 2.14h and i). From the previous discussions, it is seen that the axial deformations (D_1 and D_7) depend only on the axial forces (P_1 and P_7). Similarly, the torsional deformations (D_4 and D_{10}) depend only on the torsional moments (P_4 and P_{10}).

If the XYZ axes are chosen arbitrarily, then the bending deformations in the XY plane (D_2, D_6, D_8, and D_{12}) not only depends the bending forces the XY plane (P_2, P_6, P_8, and P_{12}) but also on the bending forces acting the XZ plane (P_3, P_5, P_9, and P_{11}). However, if the XY plane and XZ plane coincides with the principal axes of the cross section, then the bending displacements and forces in the two planes are independent of each other.

In this book, the centroidal axis of the element is chosen as the X-axis. The major principal axis of the section is chosen as the Z-axis and the minor principal axis is taken as the Y-axis (Figure 2.14h).

The stiffness matrix of the space frame element is obtained using the superposition principle (Figure 2.14i).

Thus, the stiffness matrix is

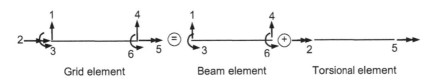

Grid element   Beam element   Torsional element

FIGURE 2.14g
Superposition principle for plane frame element.

FIGURE 2.14h
Coordinate axes for elements.

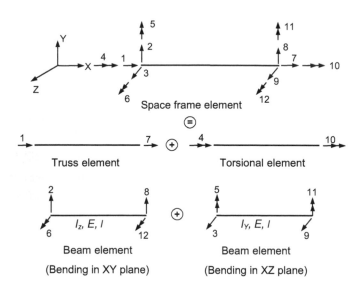

FIGURE 2.14i
Superposition principle for space frame element.

$$[K] = \begin{bmatrix}
EA/l & 0 & 0 & 0 & 0 & 0 \\
0 & 12EI_Z/l^3 & 0 & 0 & 0 & 6EI_Z/l^2 \\
0 & 0 & 12EI_Y/l^3 & 0 & -6EI_Y/l^2 & 0 \\
0 & 0 & 0 & GJ/l & 0 & 0 \\
0 & 0 & -6EI_Y/l^2 & 0 & 4EI_Y/l & 0 \\
0 & 6EI_Z/l^2 & 0 & 0 & 0 & 4EI_Z/l \\
-EA/l & 0 & 0 & 0 & 0 & 0 \\
0 & -12EI_Z/l^3 & 0 & 0 & 0 & -6EI_Z/l^2 \\
0 & 0 & -12EI_Y/l^3 & 0 & 6EI_Y/l^2 & 0 \\
0 & 0 & 0 & -GJ/l & 0 & 0 \\
0 & 0 & -6EI_Y/l^2 & 0 & 2EI_Y/l & 0 \\
0 & 6EI_Z/l^2 & 0 & 0 & 0 & 2EI_Z/l
\end{bmatrix}$$

$$\begin{bmatrix}
-EA/l & 0 & 0 & 0 & 0 & 0 \\
0 & -12EI_Z/l^3 & 0 & 0 & 0 & 6EI_Z/l^2 \\
0 & 0 & -12EI_Y/l^3 & 0 & -6EI_Y/l^2 & 0 \\
0 & 0 & 0 & -GJ/l & 0 & 0 \\
0 & 0 & 6EI_Y/l^2 & 0 & 2EI_Y/l & 0 \\
0 & -6EI_Z/l^2 & 0 & 0 & 0 & 2EI_Z/l \\
EA/l & 0 & 0 & 0 & 0 & 0 \\
0 & 12EI_Z/l^3 & 0 & 0 & 0 & -6EI_Z/l^2 \\
0 & 0 & 12EI_Y/l^3 & 0 & 6EI_Y/l^2 & 0 \\
0 & 0 & 0 & GJ/l & 0 & 0 \\
0 & 0 & 6EI_Y/l^2 & 0 & 4EI_Y/l & 0 \\
0 & -6EI_Z/l^2 & 0 & 0 & 0 & 4EI_Z/l
\end{bmatrix}$$

Comments

The stiffness matrixes of the above elements are singular, i.e., the matrixes are not invertible. If one row or column of the matrix can be written as a linear combination of other rows or columns, then the determinant of the matrix is zero. Using this property of the matrix, it can be seen that the determinant of the stiffness matrices of the elements are zero. Hence, inverse does not exist. In other words, the flexibility matrix does not exist for these elements.

This inference can be also obtained by observing the fact that the rigid body movements of the elements are not prevented. Hence, the elements are not stable structures. For getting the flexibility coefficients, it is necessary to apply unit load at a coordinate and find the displacements. An unstable structure cannot resist any load. Hence, for an unstable structure, even though the stiffness matrix can be developed, flexibility matrix does not exist.

Example 2.3

Determine the stiffness matrix of the structures shown in Figure 2.15a.

Solution

Structure 1

The deformed shape of the structure, the FBD of members, and joints are shown in Figure 2.15b. In the figure, $l_{AB} = 5$ m and $l_{BC} = 6$ m.

From FBD of joint,

$$K_{11} = \frac{4EI}{l_{AB}} + \frac{4EI}{l_{BC}} = \frac{22}{15}EI$$

Structure 2

- Elements of first column ($D_1 = 1$, $D_2 = D_3 = 0$) (Figure 2.15c)
  From the FBD of joints,

$$K_{11} = 37.5\,EI/l^3,\ K_{21} = 3\,EI/l^2,\ K_{31} = 1.5\,EI/l^2$$

- Elements of second column ($D_2 = 1$, $D_1 = D_3 = 0$)
  Using the FBD of joints,

$$K_{12} = 3\,EI/l^2,\ K_{22} = 8\,EI/l,\ K_{32} = 2\,EI/l$$

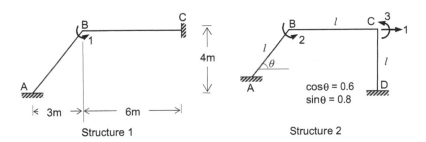

Structure 1   Structure 2

FIGURE 2.15a
Example 2.3.

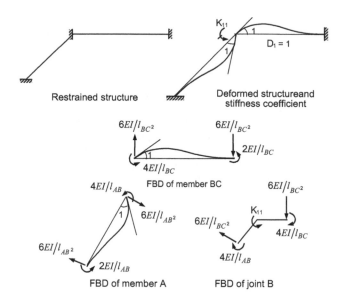

FIGURE 2.15b
Restrained structure and stiffness coefficient.

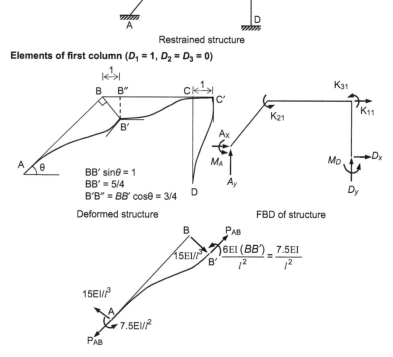

FIGURE 2.15c
Deformed structure and stiffness coefficient.

(Continued)

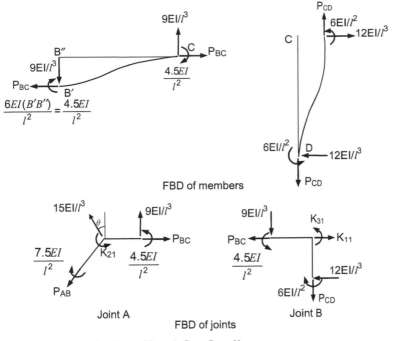

FBD of members

Joint A

FBD of joints

Joint B

Elements of second column (D_2 = 1, D_1 = D_3 = 0)

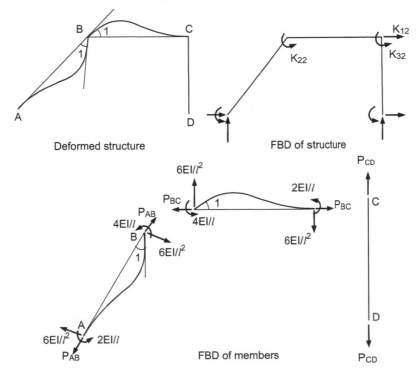

Deformed structure

FBD of structure

FBD of members

FIGURE 2.15c (CONTINUED)
Deformed structure and stiffness coefficient.

(Continued)

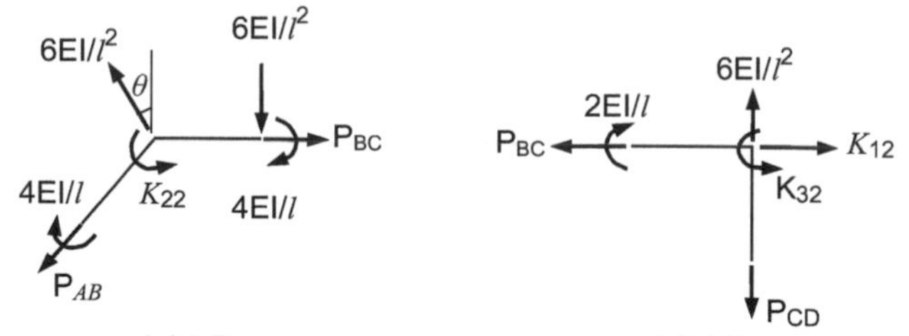

Joint B

Joint C

FBD of joints

Elements of third column ($D_3 = 1$, $D_1 = D_2 = 0$)

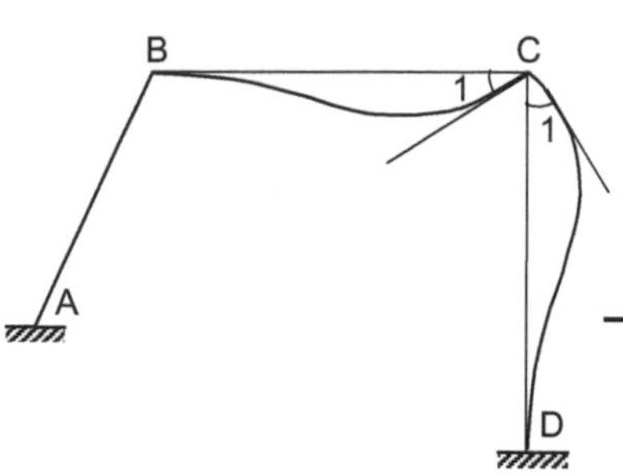

Deformed structure

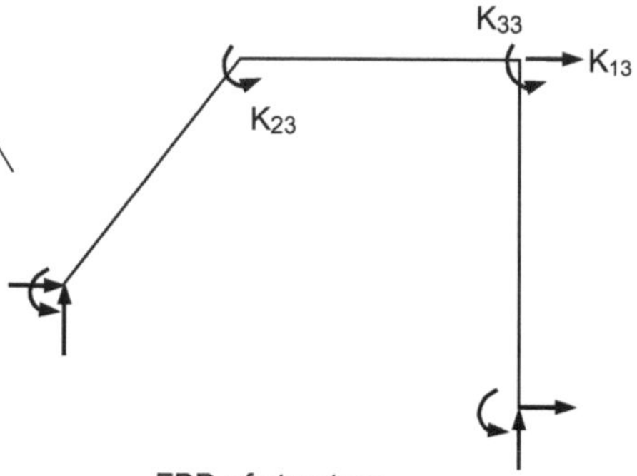

FBD of structure

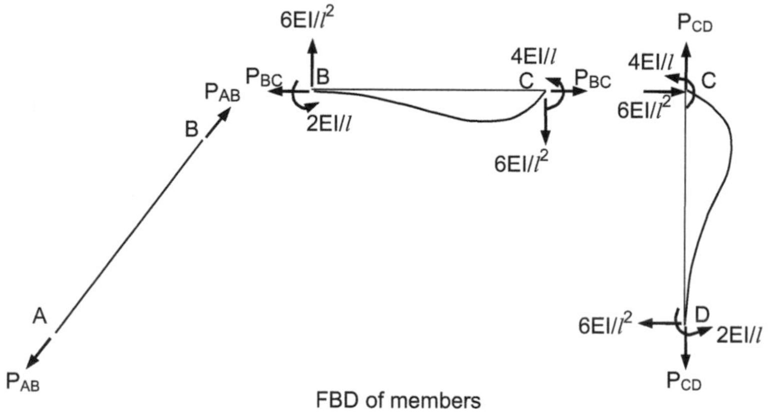

FBD of members

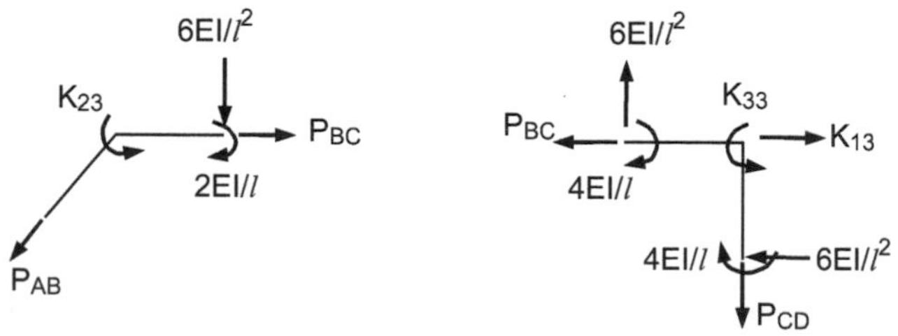

Joint B

Joint C

FBD of joints

FIGURE 2.15c (CONTINUED)
Deformed structure and stiffness coefficient.

- Elements of third column ($D_3 = 1$, $D_1 = D_2 = 0$)

$$K_{13} = 1.5EI/l^2, \ K_{23} = 2EI/l, \ K_{33} = 8EI/l$$

- Stiffness matrix

$$[K] = EI \begin{bmatrix} 37.5/l^3 & 3/l^2 & 1.5/l^2 \\ 3/l^2 & 8/l & 2/l \\ 1.5/l^2 & 2/l & 8/l \end{bmatrix}$$

2.3 Work and Energy

2.3.1 Work W

The work dW of a force P acting through a change in displacement dD in the direction of the force is defined as the product $P \, dD$. For the total displacement D_m, the total work W is

$$W = \int_0^{D_m} P \, dD \tag{2.11}$$

Equation 2.11 represents the area between the force–displacement curve and the displacement axis (Figure 2.16a).

When the force is constant over the full range of displacement (Figure 2.16b), then the work is

$$W = PD \tag{2.12}$$

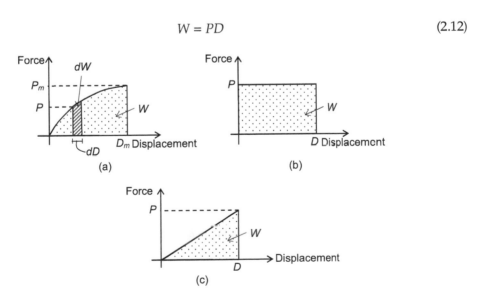

FIGURE 2.16
Work.

For a linear force–displacement relation (Figure 2.16c), the work is

$$W = \frac{1}{2}PD \tag{2.13}$$

The above equations are applicable for a structure having a single coordinate. For a structure having n coordinates, the total work is

$$W = \sum_{i=1}^{n} \int_{0}^{D_i} P_i \, dD_i \tag{2.14}$$

When the force is constant, then

$$W = P_1 D_1 + P_2 D_2 + \cdots + P_n D_n = \sum_{i=1}^{n} P_i D_i = \{P\}^T \{D\} \tag{2.15}$$

where $\{P\}^T$ is the transpose of the force vector $\{P\}$ and $\{D\}$ is the displacement vector.

For a linear force–displacement relation,

$$W = \frac{1}{2}P_1 D_1 + \frac{1}{2}P_2 D_2 + \cdots + \frac{1}{2}P_n D_n = \frac{1}{2}\sum_{i=1}^{n} P_i D_i = \frac{1}{2}\{P\}^T \{D\} \tag{2.16}$$

2.3.2 Complementary Work W^*

The increment of complementary work dW^* is the work done by the incremental force dP over a displacement D. Thus

$$dW^* = D \, dP$$

For the total force P_m, the complementary work W^* is

$$W^* = \int_{0}^{P_m} D \, dP \tag{2.17}$$

Equation 2.17 is the area between the force–displacement curve and the force axis (Figure 2.17a).

If the displacement is constant over the full range of force (Figure 2.17b), then

$$W^* = DP \tag{2.18}$$

For a linear force–displacement relation (Figure 2.17c),

$$W^* = \frac{1}{2}DP \tag{2.19}$$

For a system having n coordinates, the total complementary work W^* is

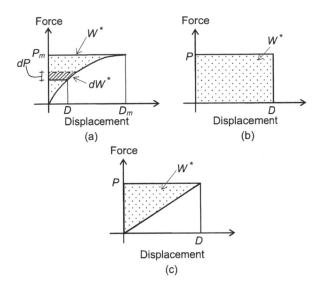

FIGURE 2.17
Complementary work.

$$W^* = \sum_{i=1}^{n} \int_0^{P_i} D_i \, dP_i \tag{2.20}$$

If the displacement is constant over the full range of forces, then

$$W^* = \sum_{i=1}^{n} D_i P_i = \{D\}^T \{P\} \tag{2.21}$$

For a linear force–displacement relationship,

$$W^* = \frac{1}{2}\sum_{i=1}^{n} D_i P_i = \frac{1}{2}\{D\}^T \{P\} \tag{2.22}$$

where $\{D\}^T$ is the transpose of the displacement vector.

Comparing Eqs. 2.16 and 2.22, it can be seen that for a linear force–displacement relationship,

$$W = W^* \tag{2.23}$$

The above relation can also be also obtained from Figures 2.16c and 2.17c.

2.3.3 Strain Energy *U*

Consider a linear elastic solid subject to loads. The stresses acting on an elemental cube of volume dV is shown in Figure 2.18. The independent stress components and the corresponding strains are represented by the column vectors $\{\sigma\}$ and $\{\varepsilon\}$. Thus, we have

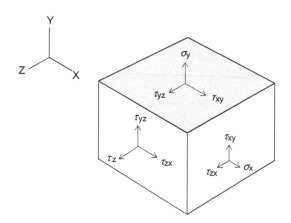

FIGURE 2.18
Stress acting on an infinitesimal element.

$$\{\sigma\} = \begin{Bmatrix} \sigma_X \\ \sigma_Y \\ \sigma_Z \\ \tau_{XY} \\ \tau_{YZ} \\ \tau_{ZX} \end{Bmatrix}, \quad \{\varepsilon\} = \begin{Bmatrix} \varepsilon_X \\ \varepsilon_Y \\ \varepsilon_Z \\ \gamma_{XY} \\ \gamma_{YZ} \\ \gamma_{ZX} \end{Bmatrix}$$

The strain energy U is defined as the internal work of stresses acting through incremental strains integrated over the total strains and over the volume. Hence,

$$U = \sum_{i=1}^{n_s} \int_V \left(\int_0^{\varepsilon_i} \sigma \, d\varepsilon \right) dv \tag{2.24}$$

where n_s is the number of strain components and V is the volume of the linear elastic solid.
$\int_0^{\varepsilon_i} \sigma \, d\varepsilon$ is the area between the stress–strain curve and the strain axis (Figure 2.19a) and is known as the *strain energy density* $\bar{U}$ (i.e., $\bar{U}$ is strain energy per unit volume). For a linear stress–strain curve, $\bar{U}$ is $1/2\sigma_i\varepsilon_i$ (Figure 2.19c). Hence, Eq. 2.24 becomes

$$U = \frac{1}{2} \sum_{i=1}^{n_s} \int_V \sigma_i \varepsilon_i dv = \frac{1}{2} \int_V \{\sigma\}^T \{\varepsilon\} dV \tag{2.25}$$

where $\{\sigma\}^T$ is the transpose of the stress vector.

2.3.4 Complementary Strain Energy U^*

The complementary strain energy U^* is defined as the internal work of strains acting through incremental stress integrated over the total stresses and over the volume. Thus, we have

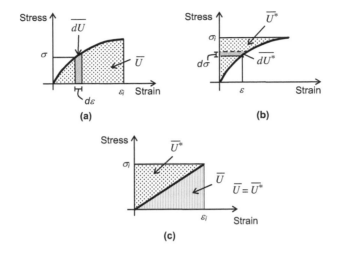

FIGURE 2.19
Strain energy and complementary strain energy: (a) strain energy density $\bar{U}$, (b) complementary strain energy density $\bar{U}^*$, and (c) $\bar{U}$ and $\bar{U}^*$ for linearly elastic.

$$U^* = \sum_{i=1}^{n_s} \int_V \left(\int_0^{\sigma_i} \varepsilon \, d\sigma \right) dv \qquad (2.26)$$

$\int_0^{\sigma_i} \varepsilon \, d\sigma$ is the area between the stress–strain curve and the stress axis (Figure 2.19b) and is known as the *complementary strain energy density* $\bar{U}^*$ ($\bar{U}^*$ is complementary strain energy per unit volume). In the case of linearly elastic materials (σ–ε relation is linear), $\bar{U}^*$ is $1/2\varepsilon_i\sigma_i$ (Figure 2.19c). Hence,

$$U^* = \frac{1}{2}\sum_{i=1}^{n_s} \int_V \varepsilon_i \sigma_i \, dv = \frac{1}{2}\int_V \{\varepsilon\}^T \{\sigma\} dV \qquad (2.27)$$

Comparing Eqs. 2.25 and 2.27, it is observed that for linearly elastic materials,

$$U = U^* \qquad (2.28)$$

2.3.5 Law of Conservation of Energy

According to the principle of conservation of energy, the work done by the loads W is equal to the strain energy U stored in the structure. Hence,

$$W = U \qquad (2.29)$$

Similarly, the principle of conservation of complementary energy leads to the condition

$$W^* = U^* \qquad (2.30)$$

2.3.5.1 Strain Energy in Terms of Flexibility and Stiffness Matrices

For linearly elastic structures, Eqs. 2.29 and 2.30 take the form

$$U = W = \frac{1}{2}\{P\}^T \{D\} \tag{2.31}$$

Also,

$$U = U^* = W^* = \frac{1}{2}\{D\}^T \{P\} \tag{2.32}$$

But $\{D\} = [F] \{P\}$ and $\{P\} = [K] \{D\}$ (from Eqs. 2.5 and 2.7)

Thus, we have

$$U = \frac{1}{2}\{P\}^T \{D\} = \frac{1}{2}\{P\}^T [F]\{P\} \tag{2.33}$$

or

$$U = \frac{1}{2}\{D\}^T \{P\} = \frac{1}{2}\{D\}^T [K]\{D\} \tag{2.34}$$

2.4 Symmetry of Flexibility and Stiffness Matrices

From Eq. 2.31,

$$U = \frac{1}{2}\{P\}^T \{D\}$$

But

$$\{P\} = [K]\{D\}$$

$$\Rightarrow \quad \{P\}^T = \{D\}^T [K]^T$$

Hence,

$$U = \frac{1}{2}\{D\}^T [K]^T \{D\} \tag{2.35}$$

Comparing Eqs. 2.34 and 2.35 gives

$$[K] = [K]^T \tag{2.36}$$

$$\Rightarrow \quad K_{ij} = K_{ji}$$

Thus, the stiffness matrix of a structure is symmetric. Similarly from Eq. 2.32,

$$U = \frac{1}{2}\{D\}^T\{P\}$$

Since

$$\{D\} = [F]\{P\}$$

$$\Rightarrow \quad \{D\}^T = \{P\}^T[F]^T$$

Therefore,

$$U = \frac{1}{2}\{P\}^T[F]^T\{P\} \tag{2.37}$$

From Eqs. 2.33 and 2.37, it can be seen that

$$[F] = [F]^T$$

$$\Rightarrow \quad F_{ij} = F_{ji}$$

Hence, the flexibility matrix of a structure is also symmetric.

2.5 Relation between Stiffness and Flexibility Coefficients and Strain Energy

Consider the structure with three coordinates shown in Figure 2.3. The strain energy U for the structure using Eq. 2.34 is

$$U = \frac{1}{2}\{D\}^T[K]\{D\}$$

$$= \frac{1}{2}\begin{bmatrix} D_1 & D_2 & D_3 \end{bmatrix} \begin{bmatrix} K_{11} & K_{12} & K_{13} \\ K_{21} & K_{22} & K_{23} \\ K_{31} & K_{32} & K_{33} \end{bmatrix} \begin{Bmatrix} D_1 \\ D_2 \\ D_3 \end{Bmatrix}$$

$$= \frac{1}{2}[K_{11}D_1^2 + K_{12}D_1D_2 + K_{13}D_1D_3 + K_{21}D_2D_1 + K_{22}D_2^2 + K_{23}D_2D_3$$

$$+ K_{31}D_3D_1 + K_{32}D_3D_2 + K_{33}D_3^2]$$

The partial derivative of U with respect to the displacements gives

$$\frac{\partial U}{\partial D_1} = \frac{1}{2}(2K_{11}D_1 + K_{12}D_2 + K_{13}D_3 + K_{21}D_2 + K_{31}D_3)$$

Since $[K]$ is symmetric, $K_{12} = K_{21}$ and $K_{13} = K_{31}$

$$\Rightarrow \quad \frac{\partial U}{\partial D_1} = K_{11}D_1 + K_{12}D_2 + K_{13}D_3$$

Similarly,

$$\frac{\partial U}{\partial D_2} = K_{21}D_1 + K_{22}D_2 + K_{23}D_3 \tag{g}$$

$$\frac{\partial U}{\partial D_3} = K_{31}D_1 + K_{32}D_2 + K_{33}D_3$$

Differentiating once again with respect to the displacements gives

$$\frac{\partial^2 U}{\partial D_1^2} = K_{11}, \quad \frac{\partial^2 U}{\partial D_1\,\partial D_2} = \frac{\partial^2 U}{\partial D_2\,\partial D_1} = K_{12} = K_{21},$$

$$\frac{\partial^2 U}{\partial D_2^2} = K_{22}, \quad \frac{\partial^2 U}{\partial D_3^2} = K_{33}, \quad \frac{\partial^2 U}{\partial D_1\,\partial D_3} = \frac{\partial^2 U}{\partial D_3\,\partial D_1} = K_{13} = K_{31} \tag{h}$$

$$\frac{\partial^2 U}{\partial D_2\,\partial D_3} = \frac{\partial^2 U}{\partial D_3\,\partial D_2} = K_{23} = K_{32}$$

For an n coordinate structure, Eq. (g) becomes

$$\frac{\partial U}{\partial D_i} = \sum_{j=1}^{n} K_{ij}D_j \tag{2.39}$$

But as per Eq. 2.6b, $\displaystyle\sum_{j=1}^{n} K_{ij}D_j = P_i$. Hence,

$$\frac{\partial U}{\partial D_i} = P_i \tag{2.40}$$

which is the Castigliano's first theorem.

Equation (h) for a structure having n coordinate is

$$\frac{\partial^2 U}{\partial D_i\,\partial D_j} = \frac{\partial^2 U}{\partial D_j\,\partial D_i} = K_{ij} \tag{2.41}$$

Thus, the stiffness coefficient K_{ij} is equal to the second partial derivative of strain energy U with respect to the displacements D_i and D_j.

Similarly using Eq. 2.33, it can be proved that

$$\frac{\partial U}{\partial P_i} = \sum_{j=1}^{n} F_{ij}D_j \tag{2.42}$$

Comparing with Eq. 2.4b gives

$$\frac{\partial U}{\partial P_i} = D_i \qquad (2.43)$$

which is the Castigliano's second theorem.

Differentiating Eq. 2.42 with respect to the force P_j gives

$$\frac{\partial^2 U}{\partial P_i \partial P_j} = F_{ij} \qquad (2.44)$$

Thus, the flexibility coefficient F_{ij} is equal to the second partial derivative of strain energy U with respect to the forces P_i and P_j.

Example 2.4

Develop the stiffness matrix of the truss and beam elements using Eq. 2.41.

Solution

Truss Element

- Strain energy
  In the rod subjected to axial force P shown in Figure 2.20a,

$$\sigma_x = P/A$$

$$\varepsilon_x = \sigma_x/E = P/AE$$

Hence, the incremental strain energy dU is

$$dU = \frac{1}{2}\sigma_x \varepsilon_x dV = \frac{1}{2}\frac{P^2}{A^2 E}dV$$

But $dV = dA\,dx$

$$\Rightarrow \quad dU = \frac{1}{2}\frac{P^2}{A^2 E}dA\,dx$$

$$U = \int_V dU = \frac{1}{2}\int_l \frac{P^2}{A^2 E}\left(\int_A dA\right)dx = \frac{1}{2}\int_l \frac{P^2}{AE}dx \quad \left[\because \int_A dA = A\right]$$

FIGURE 2.20a
Stress and strain in a rod subjected to axial tension.

For a prismatic element with constant axial force,

$$U = \frac{P^2 l}{2AE} \tag{i}$$

Let u be the displacement in the truss element at a distance x from a (Figure 2.20b). Then,

$$\varepsilon_x = \frac{du}{dx} = u'$$

$$\sigma_x = E\varepsilon_x$$

$$\therefore \quad dU = \frac{1}{2}\sigma_x \varepsilon_x = \frac{1}{2}E\varepsilon_x^2 = \frac{1}{2}E(u')^2$$

$$\Rightarrow \quad U = \frac{1}{2}\int_V E(u')^2 dV = \frac{1}{2}\int_l E(u')^2 \left(\int_A dA\right) dx = \frac{1}{2}\int_l AE(u')^2 dx \tag{j}$$

The displacement u is assumed to have a linear variation

$$\Rightarrow \quad u = a_0 + a_1 x$$

The constants a_0 and a_1 are determined using the conditions, at $x = 0$, $u = D_1$, and at $x = l$, $u = D_2$

Hence, $a_0 = D_1$ and $a_1 = (D_2 - D_1)/l$

Therefore, $u = D_1 + \dfrac{(D_2 - D_1)x}{l}$

$$\Rightarrow \quad u' = \frac{(D_2 - D_1)}{l}$$

Using Eq. (j),

$$\text{Strain energy } U = \frac{1}{2}\int_l AE(u')^2 dx = \frac{AE}{2}\int_0^l \left(\frac{D_2 - D_1}{l}\right)^2 dx$$

$$\Rightarrow \quad U = \frac{AE}{2l}\left(D_1^2 + D_2^2 - 2D_1 D_2\right)$$

- Stiffness coefficients

$$K_{11} = \frac{\partial^2 U}{\partial D_1^2} = \frac{AE}{l}, \; K_{12} = K_{21} = \frac{\partial^2 U}{\partial D_1 \partial D_2} = -\frac{AE}{l}, \; K_{22} = \frac{\partial^2 U}{\partial D_2^2} = \frac{AE}{l}$$

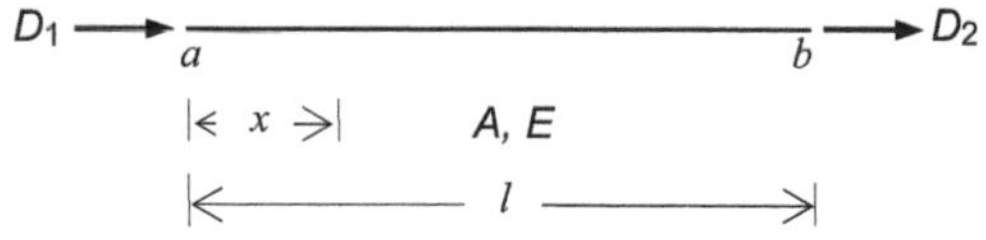

FIGURE 2.20b
End displacements in a truss element.

- Stiffness matrix

$$[K] = \frac{AE}{l} \begin{bmatrix} 1 & -1 \\ -1 & 1 \end{bmatrix}$$

Beam Element

- Strain energy
  In a beam, the normal stress σ_x and normal strain ε_y at a distance y from the neutral axis is (Figure 2.20c),

$$\sigma_x = -\frac{M}{I}y \text{ and } \varepsilon_x = -\kappa y$$

where $\kappa = $ curvature $\approx d^2v/dx^2 = v''$
$v = $ transverse displacement of beam
The incremental strain energy dU is

$$dU = \frac{1}{2}\sigma_x \varepsilon_x dV$$

But $\varepsilon_x = \sigma_x/E$

$$\Rightarrow dU = \frac{1}{2}\frac{\sigma_x^2}{E}dV = \frac{1}{2}\frac{M^2}{I^2 E}y^2 dA dx \quad (\because dV = dA dx)$$

Strain energy $U = \int dU = \frac{1}{2}\int_l \frac{M^2}{I^2 E}\left(\int_A y^2 \, dA\right)dx$

But $\int_A y^2 \, dA = I$

$$\Rightarrow U = \frac{1}{2}\int_l \frac{M^2}{EI}dx \tag{k}$$

For a prismatic beam subjected to constant bending moment,

$$U = \frac{M^2 l}{2EI} \tag{l}$$

Alternatively,

$$dU = \frac{1}{2}\sigma_x \varepsilon_x dV = \frac{1}{2}E\varepsilon_x^2 dV \quad (\because \sigma_x = E\varepsilon_x)$$

$$= \frac{1}{2}Ey^2 (v'')^2 \, dV \quad (\because \varepsilon_x = -yv'')$$

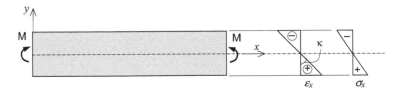

FIGURE 2.20c
Stress and strain in a beam subjected to bending moment.

Therefore, $U = \int_V dU = \dfrac{1}{2}\int_V Ey^2\,(v'')^2\,dA\,dx = \dfrac{1}{2}\int_l E(v'')^2\left(\int_A y^2\,dA\right)dx$

$$\Rightarrow \quad U = \dfrac{1}{2}\int_l EI(v'')^2\,dx \qquad\qquad\text{(m)}$$

The transverse displacement of the beam element is of the form

$$v = a_0 + a_1 x + a_2 x^2 + a_3 x^3$$

where $a_0,\,a_1,\,a_2,$ and a_3 are determined using the conditions (Figure 2.20d),
At $x = 0$, $v = D_1$; at $x = l$, $v = D_3$; at $x = 0$, $dv/dx = D_2$ and at $x = l$, $dv/dx = D_4$.
Thus, we have

$$v = \left(1 - 3\xi^2 + 2\xi^3\right)D_1 + l\left(\xi - 2\xi^2 + \xi^3\right)D_2 + \left(3\xi^2 - 2\xi^3\right)D_3 + l\left(-\xi^2 + \xi^3\right)D_4$$

where $\xi = x/l$
Using Eq. m, the strain energy stored in the beam element is

$$U = \dfrac{EI}{2l^3}\left(\begin{array}{l} 12D_1^2 + 4D_2^2 l^2 + 12D_3^2 + 4D_4^2 l^2 + 12D_1 D_2 l - 24D_1 D_3 + 12D_1 D_4 l \\ -12D_2 D_3 l + 4D_2 D_4 l^2 - 12D_3 D_4 l \end{array}\right)$$

- Stiffness coefficients

$$K_{11} = \dfrac{\partial^2 U}{\partial D_1^2} = 12EI/l^3,\ K_{12} = K_{21} = \dfrac{\partial^2 U}{\partial D_1\,\partial D_2} = -6EI/l^2$$

$$K_{13} = K_{31} = \dfrac{\partial^2 U}{\partial D_1\,\partial D_3} = -12EI/l^3$$

$$K_{14} = K_{41} = \dfrac{\partial^2 U}{\partial D_1\,\partial D_4} = 6EI/l^2$$

$$K_{22} = \dfrac{\partial^2 U}{\partial D_2^2} = 4EI/l,\ K_{23} = K_{32} = \dfrac{\partial^2 U}{\partial D_1\,\partial D_3} = -6EI/l^2$$

$$K_{24} = K_{42} = \dfrac{\partial^2 U}{\partial D_2\,\partial D_4} = 2EI/l,\ K_{33} = \dfrac{\partial^2 U}{\partial D_3^2} = 12EI/l^3$$

$$K_{34} = K_{43} = \dfrac{\partial^2 U}{\partial D_3\,\partial D_4} = -6EI/l^2,\ K_{44} = \dfrac{\partial^2 U}{\partial D_4^2} = 4EI/l$$

FIGURE 2.20d
End displacements in a beam element.

- Stiffness matrix

$$[K] = EI \begin{bmatrix} 12/l^3 & 6/l^2 & -12/l^3 & 6/l^2 \\ 6/l^2 & 4/l & -6/l^2 & 2/l \\ -12/l^3 & -6/l^2 & 12/l^3 & -6/l^2 \\ 6/l^2 & 2/l & -6/l^2 & 4/l \end{bmatrix}$$

Example 2.5

Develop the flexibility matrix of the structures shown in Figure 2.21a using Eq. 2.44.

Solution

Structure 1

- Strain energy U

    Let P_1 be the axial force acting at coordinate 1. The FBD of structure is shown in Figure 2.21b

    The axial force at a section x–x is P_1. Then the strain energy of the structure is (Eq. i, Example 2.4)

$$U = \frac{P^2 l}{2AE} = \frac{P_1^2 l}{2AE}$$

- Flexibility coefficient and flexibility matrix

$$F_{11} = \frac{\partial^2 U}{\partial P_1^2} = \frac{l}{AE}$$

Hence, the flexibility of the structure is $F = l/AE$.

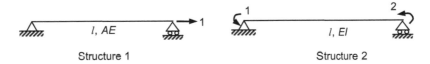

FIGURE 2.21a
Example 2.5.

FIGURE 2.21b
FBD of Structure 1.

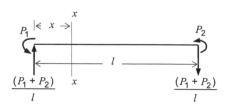

FIGURE 2.21c
FBD of Structure 2.

Structure 2

- Strain energy U
  The forces acting at the two coordinates are P_1 and P_2, respectively. The FBD of the structure is shown in Figure 2.21c.
  The bending moment at section x–x is

$$M = \frac{(P_1 + P_2)}{l}x - P_1$$

The strain energy of the structure (Eq. k in Example 2.4) is

$$U = \frac{1}{2}\int_0^l \frac{M^2 dx}{2EI} = \frac{1}{2EI}\left[\left(P_1^2 + P_2^2 + 2P_1P_2\right)\frac{l}{3} + P_1^2 l - \left(P_1^2 + P_1P_2\right)l\right]$$

- Flexibility coefficients

$$F_{11} = \frac{\partial^2 U}{\partial P_1^2} = l/3EI$$

$$F_{12} = F_{21} = \frac{\partial^2 U}{\partial P_1\,\partial P_2} = -l/6EI,$$

$$F_{22} = \frac{\partial^2 U}{\partial P_2^2} = l/3EI$$

- Flexibility matrix

$$[F] = \frac{l}{6EI}\begin{bmatrix} 2 & -1 \\ -1 & 2 \end{bmatrix}$$

Problems

2.1 Determine the flexibility and stiffness matrices of the structures shown in Figure 2.22.
2.2 Determine the stiffness matrices of the structures shown in Figure 2.23.

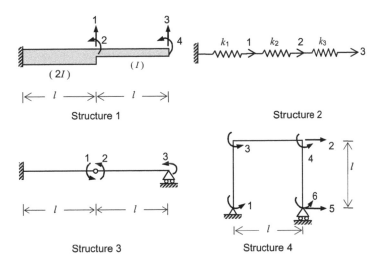

FIGURE 2.22
Problem 2.1.

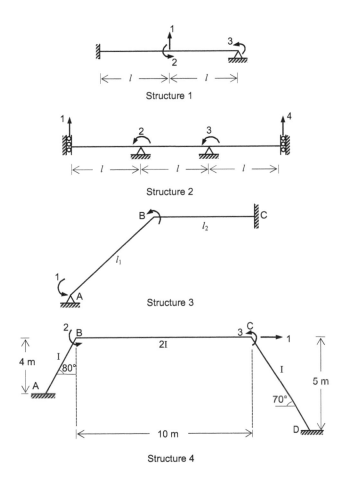

FIGURE 2.23
Problem 2.2.

3

Flexibility Method

3.1 Introduction

For a designer, an analysis procedure that directly gives forces in a structure will be very useful. This is due to the fact that the proportions of the structure are usually based on forces rather on displacements. The analysis approach in which equations are written in terms of unknown forces is known as the *flexibility method*. Since in this method, the unknowns are forces, this method is also known as the *force method*. In the flexibility method, equations to find unknown forces (in the case of statically indeterminate structures) are obtained by forcing the calculated displacements at the redundant locations to be consistent with the requirements of compatibility. Hence, this approach is also known as the *method of consistent deformation*.

In the flexibility method, to begin with, equations of equilibrium are written for the forces in the structure and the elements. These equations alone are sufficient to find the unknown forces in a statically determinate structure. However, additional equations are necessary to find forces in a statically indeterminate structure. For this purpose, the force–displacement relations are written using flexibility coefficients and compatibility conditions are used to find the redundant forces. The order in which equations are used in the flexibility method is shown in Figure 3.1.

In this chapter, procedures required to analyze both statically determinate and indeterminate structures using the flexibility method are discussed. The procedures are explained by considering the analysis of planar structures.

3.2 Coordinates for Forces and Displacements

In the flexibility (or force) method of analysis, importance is given to forces. Hence, the coordinates are chosen based on forces. Once the coordinates are defined, they also indicate the direction of corresponding displacements.

3.2.1 Global Coordinates: Action and Redundant Coordinates

In the flexibility method, the global coordinates at the nodes indicate the direction of forces $\{P\}$ and corresponding displacements $\{D\}$ of the structure. For the analysis of a statically indeterminate structure using the flexibility method, it is necessary to identify the

Matrix Methods of Structural Analysis

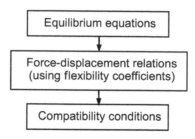

FIGURE 3.1
Equations used in the flexibility method.

redundant forces so as to develop the primary structure. After arriving at the primary structure, it is subjected to loads and redundant forces. Hence, two types of global coordinates are required: one to define loads and the other for redundant forces. The global coordinates that identify the loads or actions are called the *action coordinates* and the global coordinates that indicate the redundant forces are called the *redundant coordinates*. The structural analyst has the freedom to choose different types of primary structure for the given indeterminate structure (see Figure 1.16). Hence, the redundant coordinates will in fact depend on the type of primary structure chosen. However, the number of redundant coordinates in any case will be equal to DSI of the structure.

Thus in the flexibility method, the force vector $\{P\}$ is partitioned into two as

$$\{P\} = \left[\begin{array}{c} \{P_A\} \\ \hline \{P_X\} \end{array} \right] \tag{3.1}$$

where $\{P_A\}$ is the *action vector* or *load vector*, which is the known vector and $\{P_X\}$ is the *redundant vector* which is unknown.

Similarly, the displacement vector $\{D\}$ is also partitioned as

$$\{D\} = \left[\begin{array}{c} \{D_A\} \\ \hline \{D_X\} \end{array} \right] \tag{3.2}$$

where $\{D_A\}$ and $\{D_X\}$ are the displacement vectors corresponding to the force vectors $\{P_A\}$ and $\{P_X\}$, respectively. The displacements corresponding to the redundant forces are known, while the displacements corresponding to the loads are unknowns. Hence, $\{D_X\}$ is the known vector and $\{D_A\}$ is the unknown vector.

In order to identify the global coordinates of a structure, it is necessary to develop the discretized structure (i.e., identity nodes and elements of a structure). After this, the action coordinates are identified at the nodes where loads are applied and where displacements are to be found out. After numbering the action coordinates, the redundant coordinates are numbered. The order of numbering the coordinates is discussed in Section 1.6. The global coordinates for some typical structures are shown in Figure 3.2.

Structure I (Figure 3.2a) is statically determinate structure. Hence, there will be no redundant coordinates. In this structure, in addition to find the element forces, it is intended to find the vertical deflection at joint *b*. Hence, two action coordinates are defined: one for vertical deflection at joint b (coordinate 1) and the other for defining the applied moment (coordinate 2). Since no load is acting at coordinate1, $P_1 = 0$.

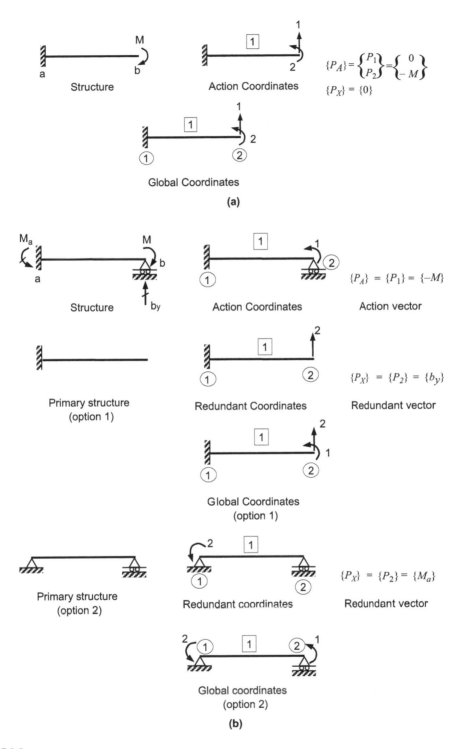

FIGURE 3.2
Global coordinates—flexibility method. (a) Structure I, (b) Structure II, (c) Structure III, (d) Structure IV, and (e) Structure V.

(Continued)

The remaining structures shown in Figure 3.2 are statically indeterminate. In the case of Structure II, after defining the action coordinates, two possible primary structures are considered. In the first one (Option 1), the vertical reaction b_y is taken as the redundant and in the second (Option 2), the reaction moment M_a is the redundant. From Figure 3.2b, it can be seen that the redundant vector $\{P_X\}$ will change according to the primary structure.

Structure III is a statically indeterminate (externally indeterminate) truss. The coordinate 3 is introduced so as to find the horizontal displacement at joint d. The reaction at support e (e_y) is taken as redundant and the global coordinates are shown in Figure 3.2c. Structure IV is a statically indeterminate (internally indeterminate) truss. The force in the member ac (P_{ac}) is chosen as the redundant and the global coordinates of the structure are indicated in Figure 3.2d.

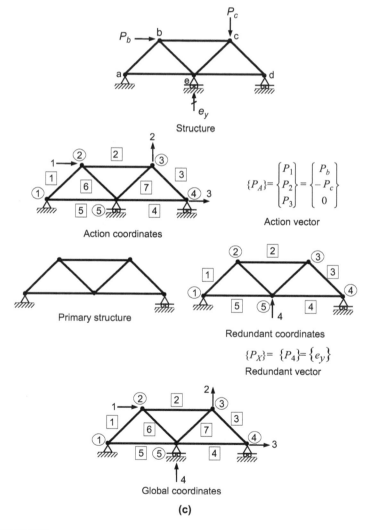

$$\{P_A\}= \begin{Bmatrix} P_1 \\ P_2 \\ P_3 \end{Bmatrix} = \begin{Bmatrix} P_b \\ -P_c \\ 0 \end{Bmatrix}$$

Action vector

$$\{P_X\}= \{P_4\}= \{e_y\}$$

Redundant vector

(c)

FIGURE 3.2 (CONTINUED)
Global coordinates—flexibility method. (a) Structure I, (b) Structure II, (c) Structure III, (d) Structure IV, and (e) Structure V.

(Continued)

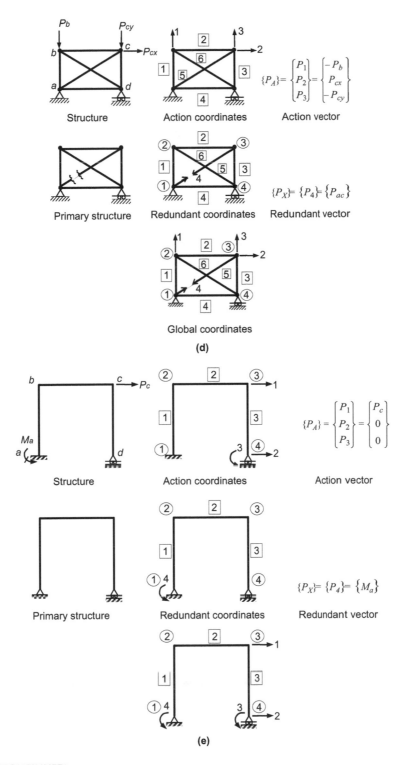

FIGURE 3.2 (CONTINUED)
Global coordinates—flexibility method. (a) Structure I, (b) Structure II, (c) Structure III, (d) Structure IV, and (e) Structure V.

In the case of plane frame structure shown in Figure 3.2e, the coordinates 2 and 3 are included so as to find the displacements at joint d. The reaction moment M_a is taken as the redundant and the global coordinates are defined accordingly.

3.2.2 Local Coordinates

The local coordinates are used to indicate forces and displacements of the elements in a structure. The details of elements used to model a planar structure analyzed using the flexibility method are given below.

3.2.2.1 Types of Elements

Figure 3.3a shows the FBD of the element in a plane frame. Let the element be the ith element of the structure. j and k are the near and far nodes of the element, respectively. x_i and y_i are the local axes of the element i, with the origin at the near node j. The three end actions (i.e., axial force N, shear force V, and bending moment M) at the nodes are also indicated in the figure.

From the FBD of the element, it can be seen that, if the three end actions at a node are known, then the three end actions at the other node can be determined using equations of equilibrium. Similarly, if the end moments of both the nodes and the axial force at any one of the node are known, the remaining end actions can be evaluated. Hence, out of the six end actions of the element, only three are the basic end actions. The three basic end actions considered in this chapter are shown in Figure 3.3b. The positive direction of the element forces and the corresponding displacements are also indicated in the figure. Thus,

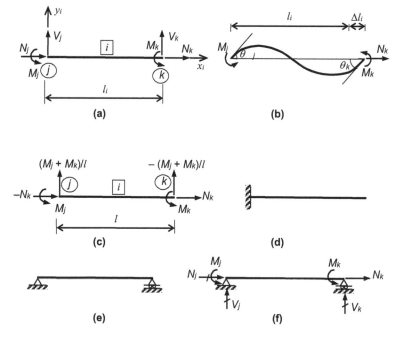

FIGURE 3.3
Plane frame element: (a) FBD of Plane frame element, (b) basic end action and corresponding basic end displacements, (c) end action, (d) basic elements I, (e) basic elements II, and (f) actions and reactions in basic element II.

the sign convention for the element forces and the corresponding displacements are as follows: counter clock wise end moments and end rotations (θ_j and θ_k) are positive, and a tensile axial force and the corresponding translation of node k (Δl_i) along the positive x_i—directions are positive quantities. The other end actions can be written in terms of basic end actions using equations of equilibrium (Figure 3.3c).

Alternatively, from Example 2.2, it is observed that to develop flexibility matrix, it is necessary to have a stable structure. Two alternatives exist to have a stable plane frame element and they are shown in Figures 3.3d and e, respectively. If the element in Figure 3.3d is chosen, then the end actions at node k (N_k, M_k, and V_k) are the basic end actions. However, if the element shown in Figure 3.3e is chosen, then the end actions M_j, M_k, and N_k are the basic end actions and the remaining end actions are the support reactions of the element (Figure 3.3f).

Thus, the different types of elements used for the analysis of two-dimensional structures using the flexibility method are shown in Figure 3.4. If both flexural and axial deformations have to be considered, then plane frame element (Figure 3.4a) is used. Beam elements (Figure 3.4b) are used when axial deformations are neglected and only flexural deformations are considered in the analysis. If only axial deformation needs to be considered, then truss element (Figure 3.4c) is used.

3.2.2.2 Element Force and Element Displacement Vectors

The local coordinates of different types of elements are shown in Figure 3.4. The local coordinates are numbered with a superscript to identify the element number. The local coordinates in the figure indicate the directions of the element forces and the corresponding element displacements. For example, in the case of the plane frame element (Figure 3.4a), the *element force vector* $\{p^i\}$ and the *element displacement vector* $\{d^i\}$ in local coordinates are

$$\{p^i\} = \begin{Bmatrix} p_1^i \\ p_2^i \\ p_3^i \end{Bmatrix} \text{ and } \{d^i\} = \begin{Bmatrix} d_1^i \\ d_2^i \\ d_3^i \end{Bmatrix} \tag{3.3}$$

The element force vector and element displacement vector of all the elements of the structure are assembled to get the *combined element force vector* $\{p\}$ and *combined element displacement vector* $\{d\}$. If there are n elements in a structure, the combined element force and displacement vectors are

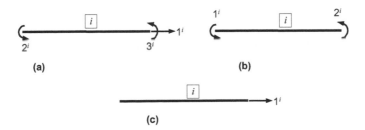

(a) (b)

(c)

FIGURE 3.4
Elements used in the flexibility method: (a) plane frame element, (b) beam element, and (c) truss element.

$$\{p\} = \begin{Bmatrix} p_1 \\ p_2 \\ p_3 \\ \vdots \\ p_n \end{Bmatrix} = \begin{Bmatrix} \{p^1\} \\ \{p^2\} \\ \vdots \\ \{p^n\} \end{Bmatrix} \text{ and } \{d\} = \begin{Bmatrix} d_1 \\ d_2 \\ d_3 \\ \vdots \end{Bmatrix} = \begin{Bmatrix} \{d^1\} \\ \{d^2\} \\ \vdots \\ \{d^n\} \end{Bmatrix} \tag{3.4}$$

To identify the combined element forces and displacements, *combined local coordinates* are used. The local coordinates of all the elements are numbered in order to get combined local coordinates. The local coordinates and combined local coordinates of the elements in typical planar structures are shown in Figure 3.5.

The local and combined local coordinates of a three-span continuous beam and plane truss are shown in Figure 3.5a and b, respectively. The plane frame shown in Figure 3.5c can be modeled using two different ways. If both axial and flexural deformations are included in analysis, then the structure is modeled using plane frame elements (Option I).

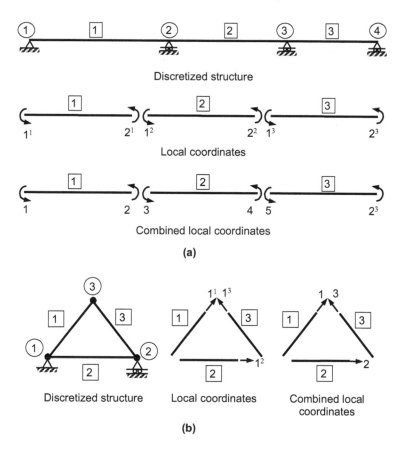

FIGURE 3.5
Local coordinates and combined local coordinates. (a) Structure I: Continuous beam, (b) Structure II: Plane truss, and (c) Structure III: Plane frame.

(Continued)

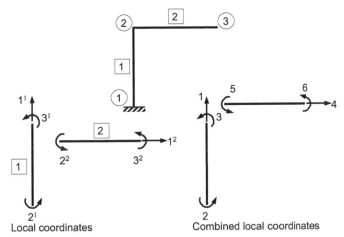

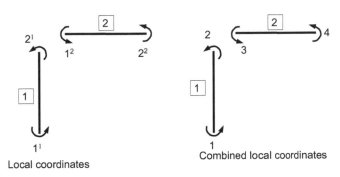

FIGURE 3.5 (CONTINUED)
Local coordinates and combined local coordinates. (a) Structure I: Continuous beam, (b) Structure II: Plane truss, and (c) Structure III: Plane frame.

However, if the axial deformations are neglected (usually done in hand computations), then the frame is modeled using beam elements (Option II). Even if the frame is analyzed using beam elements, the axial forces developed in the structure can be determined using the FBD of elements and nodes (see Example 3.17).

3.3 Equilibrium Equations and Force Transformation Matrix

The element forces $\{p\}$ are written in terms of forces acting on the structure $\{P\}$ using equations of equilibrium. Consider a structure having m combined local coordinates and n global coordinates. In this case,

$$p_1 = T_{F11}P_1 + T_{F12}P_2 + \cdots + T_{F1n}P_n$$
$$p_2 = T_{F21}P_1 + T_{F22}P_2 + \cdots + T_{F2n}P_n$$

....

....   (a)

$$p_m = T_{Fm1}P_1 + T_{Fm2}P_2 + \cdots + T_{Fmn}P_n$$

The above equations are written in matrix form as

$$\{p\} = [T_F]\{P\} \tag{3.5}$$

where
  $\{p\}$ is the combined element force vector, $\{P\}$ is the force vector, and

$$[T_F] = \begin{bmatrix} T_{F11} & T_{F12} & \cdots & T_{F1n} \\ T_{F21} & T_{F22} & \cdots & T_{F2n} \\ \cdots & \cdots & \cdots & \cdots \\ T_{Fm1} & T_{Fm2} & \cdots & T_{Fmn} \end{bmatrix}$$

is the *force transformation matrix*.

In general, a transformation matrix gives the interrelations between two coordinate systems. In this case, $[T_F]$ relates the element forces (i.e., forces in local coordinates) with forces in global coordinates using equilibrium equations. The *force transformation influence coefficient* T_{Fij} is defined as the force developed in the element corresponding to i^{th} combined local coordinate due to unit force at the jth global coordinate, while the forces in all other global coordinates are equal to zero.

3.3.1 Development of Force Transformation Matrix

Consider the force vector,

$$\{P\} = \begin{bmatrix} P_1 = 1 \\ P_2 = 0 \\ P_3 = 0 \\ \cdots \\ P_n = 0 \end{bmatrix}$$

For this load case, using Eq. a, $p_1 = T_{F11}$, $p_2 = T_{F21}$, and $p_3 = T_{F31}$, ..., $p_m = T_{Fm1}$. Thus, the combined element force vector gives the elements of first column of $[T_F]$. Hence, the elements of $[T_F]$ are obtained columnwise as shown in Figure 3.6. Thus, to get the element of jth column of $[T_F]$, a unit force is applied at the jth global coordinate and the force at all other coordinates are equal to zero. The corresponding combined element force vector gives the elements of jth column of $[T_F]$.

$$P_1=1 \quad P_2=1 \qquad P_j=1 \qquad P_n=1$$

$$[T_F] = \begin{bmatrix} T_{F11} & T_{F12} & \cdots & T_{F1j} & \cdots & T_{F1n} \\ T_{F21} & T_{F22} & \cdots & T_{F2j} & \cdots & T_{F2n} \\ T_{F31} & T_{F32} & \cdots & T_{F3j} & \cdots & T_{F3n} \\ \cdots & \cdots & \cdots & \cdots & \cdots & \cdots \\ \cdots & \cdots & \cdots & \cdots & \cdots & \cdots \\ T_{Fm1} & T_{Fm2} & & T_{Fmj} & & T_{Fmn} \end{bmatrix}$$

FIGURE 3.6
Force transformation matrix.

3.3.2 Statically Determinate Structure

In the case of statically determinate structure, since there are no redundants, $\{P\} = \{P_A\}$. Hence, the element forces $\{p\}$ are written in terms of the load vector $\{P_A\}$. Thus,

$$\{p\} = [T_{FA}]\{P_A\} \tag{3.6}$$

where $[T_{FA}]$ is the force transformation matrix indicating the influence of $\{P_A\}$. Hence, for a statically determinate structure,

$$[T_F] = [T_{FA}] \tag{3.7}$$

3.3.3 Statically Indeterminate Structure

The first step in the analysis of statically indeterminate structure is to identify the redundant vector $\{P_X\}$ and to obtain the primary structure. Then the primary structure is subjected to loads $\{P_A\}$ and $\{P_X\}$. Thus, the element forces $\{p\}$ are written in terms of $\{P_A\}$ and $\{P_X\}$ as

$$\{p\} = [T_{FA}]\{P_A\} + [T_{FX}]\{P_X\} \tag{3.8a}$$

$$= \left[\begin{array}{c|c} [T_{FA}] & [T_{FX}] \end{array} \right] \left\{ \begin{array}{c} \{P_A\} \\ \hline \{P_X\} \end{array} \right\} \tag{3.8b}$$

$$= [T_F]\{P\} \tag{3.8c}$$

where $[T_{FA}]$ and $[T_{FX}]$ are the force transformation matrices representing the influences of $\{P_A\}$ and $\{P_X\}$ on $\{p\}$, respectively. Hence, for a statically indeterminate structure,

$$[T_F] = \left[\begin{array}{c|c} [T_{FA}] & [T_{FX}] \end{array} \right] \tag{3.9}$$

Example 3.1

Develop the force transformation matrix for the statically determinate truss shown in Figure 3.7a.

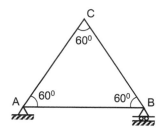

FIGURE 3.7a
Example 3.1.

Solution

The discretized structure, global, and local coordinates are shown in Figure 3.7b.
  Method I: Using equations of equilibrium
  The relation between the element forces (Figure 3.7d) and nodal forces (Figure 3.7c) can be derived using equilibrium equations.
  From Figure 3.7c,

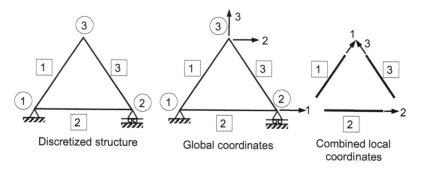

Discretized structure   Global coordinates   Combined local
                                                            coordinates

FIGURE 3.7b
Discretized structure and coordinates.

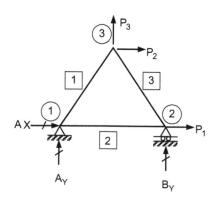

FIGURE 3.7c
Nodal forces and reactions.

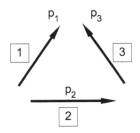

FIGURE 3.7d
Element forces.

$$\sum F_X = 0 \Rightarrow A_X = -P_1 - P_2$$

$$\sum M_1 = 0 \Rightarrow B_Y = \frac{\sqrt{3}}{2} P_2 - \frac{1}{2} P_3$$

$$\sum M_2 = 0 \Rightarrow A_Y = -\frac{\sqrt{3}}{2} P_2 - \frac{1}{2} P_3$$

The FBD of nodes 1 and 2 are shown in Figure 3.7e.
From FBD of node 1,

$$\sum F_Y = 0 \Rightarrow p_1 = -\frac{A_Y}{\sin 60} = -\frac{2}{\sqrt{3}} A_Y = -\frac{2}{\sqrt{3}} \left(-\frac{\sqrt{3}}{2} P_2 - \frac{1}{2} P_3 \right)$$

$$\Rightarrow \quad p_1 = P_2 + \frac{1}{\sqrt{3}} P_3 \qquad (a)$$

$$\sum F_X = 0 \Rightarrow p_2 = -A_X - p_1 \cos 60 = -A_X + \frac{2}{\sqrt{3}} A_Y \cos 60$$

$$= (P_1 + P_2) + \frac{\sqrt{3}}{3} \left(-\frac{\sqrt{3}}{2} P_2 - \frac{1}{2} P_3 \right)$$

$$\Rightarrow \quad p_2 = P_1 + \frac{1}{2} P_2 - \frac{\sqrt{3}}{6} P_3 \qquad (b)$$

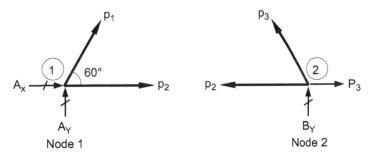

FIGURE 3.7e
FBD of nodes.

From FBD of node 2,

$$\sum F_Y = 0 \Rightarrow p_3 = -\frac{B_Y}{\sin 60} = -\frac{2}{\sqrt{3}}\left(\frac{\sqrt{3}}{2}P_2 - \frac{1}{2}P_3\right)$$

$$\Rightarrow \quad p_3 = -P_2 + \frac{1}{\sqrt{3}}P_3 \tag{c}$$

Eqs. a–c in matrix form gives

$$\{p\} = [T_{FA}]\{P\}$$

$$\begin{Bmatrix} p_1 \\ p_2 \\ p_3 \end{Bmatrix} = \begin{bmatrix} 0 & 1 & 1/\sqrt{3} \\ 1 & 1/2 & -\sqrt{3}/6 \\ 0 & -1 & 1/\sqrt{3} \end{bmatrix} \begin{Bmatrix} P_1 \\ P_2 \\ P_3 \end{Bmatrix}$$

Thus, the force transformation matrix $[T_{FA}]$ is

$$[T_{FA}] = \begin{bmatrix} 0 & 1 & 1/\sqrt{3} \\ 1 & 1/2 & -\sqrt{3}/6 \\ 0 & -1 & 1/\sqrt{3} \end{bmatrix}$$

Method II: Using Figure 3.6.

- Elements of first column of $[T_{FA}]$
  A unit load is applied at global coordinate 1 (Figure 3.7f). The forces developed in the elements are also shown in the figure and they are elements of first column of $[T_{FA}]$. In the figure, forces marked away from the node are tensile in nature and towards the node are compressive.
  The element forces are

$$\{p\} = \begin{Bmatrix} p_1 \\ p_2 \\ p_3 \end{Bmatrix} = \begin{Bmatrix} 0 \\ 1 \\ 0 \end{Bmatrix}$$

- Elements of second column of $[T_{FA}]$
  The element forces due to unit load applied at second global coordinate are

$$\{p\} = \begin{Bmatrix} p_1 \\ p_2 \\ p_3 \end{Bmatrix} = \begin{Bmatrix} 1 \\ 1/2 \\ -1 \end{Bmatrix}$$

- Elements of third column of $[T_{FA}]$
  The force in elements due to unit load acting at the third global coordinate are

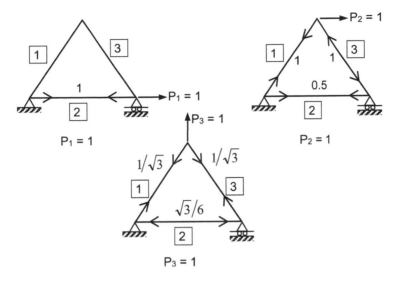

FIGURE 3.7f
Element forces due to unit load applied at global coordinates.

$$\{p\} = \left\{ \begin{array}{c} p_1 \\ p_2 \\ p_3 \end{array} \right\} = \left\{ \begin{array}{c} 1/\sqrt{3} \\ -\sqrt{3}/6 \\ 1/\sqrt{3} \end{array} \right\}$$

- Force transformation matrix $[T_{FA}]$

$$[T_{FA}] = \left[\begin{array}{ccc} 0 & 1 & 1/\sqrt{3} \\ 1 & 1/2 & -\sqrt{3}/6 \\ 0 & -1 & 1/\sqrt{3} \end{array} \right]$$

3.4 Force–Displacement Relations

The force–displacement relations both at the element and global levels are derived in this section.

3.4.1 For an Element

The force–displacement relation of the element (see Figure 3.4) is

$$\{d^i\} = [f^i]\{p^i\} \tag{3.10}$$

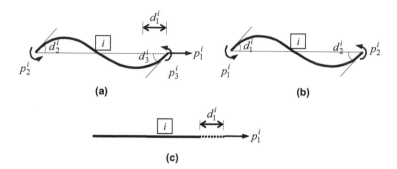

FIGURE 3.8
Forces and displacements in elements: (a) plane frame element, (b) beam element, and (c) truss element.

where $[f^i]$ is the element flexibility matrix[*].
    For the plane frame element shown in Figure 3.8a (Structure 4 of Example 2.1),

$$\left[f^i\right]=\begin{bmatrix} l/AE & 0 & 0 \\ 0 & l/3EI & -l/6EI \\ 0 & -l/6EI & l/3EI \end{bmatrix} \tag{3.11}$$

$[f^i]$ for a beam element shown in Figure 3.8b (Structure 2 of Example 2.1) is

$$[f^i]=\frac{l}{6EI}\begin{bmatrix} 2 & -1 \\ -1 & 2 \end{bmatrix} \tag{3.12}$$

For the truss element shown in Figure 3.8c (Structure 3 of Example 2.1), the element flexibility matrix is

$$\left[f^i\right]=\left[\frac{l}{AE}\right] \tag{3.13}$$

3.4.2 For the Unassembled Structure

If the structure has n elements, the force–displacement relations of the individual elements are

$$\left\{d^1\right\}=\left[f^1\right]\left\{p^1\right\}$$
$$\left\{d^2\right\}=\left[f^2\right]\left\{p^2\right\}$$
$$\vdots$$
$$\left\{d^n\right\}=\left[f^n\right]\left\{p^n\right\}$$

The above equations are written in matrix form as

---

[*] $[f^i]$ for different types of element are derived in Chapter 2, see Example 2.1.

$$\begin{Bmatrix} \{d^1\} \\ \{d^2\} \\ \vdots \\ \{d^n\} \end{Bmatrix} = \begin{bmatrix} [f^1] & 0 & 0 & \cdots & 0 \\ 0 & [f^2] & 0 & \cdots & 0 \\ \cdots & \cdots & \cdots & \cdots & \cdots \\ \cdots & \cdots & \cdots & \cdots & \cdots \\ 0 & 0 & 0 & 0 & [f^n] \end{bmatrix} \begin{Bmatrix} \{p^1\} \\ \{p^2\} \\ \vdots \\ \{p^n\} \end{Bmatrix}$$

$$\Rightarrow \quad \{d\} = [f]\{p\} \tag{3.14}$$

where $\{d\}$ and $\{p\}$ are the combined element displacement and force vectors, and

$$[f] = \begin{bmatrix} [f^1] & 0 & 0 & \cdots & 0 \\ 0 & [f^2] & 0 & \cdots & 0 \\ \cdots & \cdots & \cdots & \cdots & \cdots \\ 0 & 0 & 0 & \cdots & [f^n] \end{bmatrix}$$

is a diagonal matrix with element flexibility matrices as its constituents. Since $[f]$ is obtained by simply combining the element flexibility matrices of all the constituent elements before assembling the structure, it is known as the *unassembled flexibility matrix*.

3.4.3 For the Structure

The force–displacement relation of the structure is

$$\{D\} = [F]\{P\} \tag{3.15}$$

where $\{D\}$ and $\{P\}$ are the displacement and force vectors in global coordinates. $[F]$ is the *flexibility matrix of the structure* or the *structure flexibility matrix*.

a. Statically determinate structure

In the case of statically determinate structure, $\{D\} = \{D_A\}$ and $\{P\} = \{P_A\}$. Hence, Eq. 3.15 can be written as

$$\{D_A\} = [F_{AA}]\{P_A\} \tag{3.16}$$

Thus for a statically determinate structure, the structure flexibility matrix is

$$[F] = [F_{AA}] \tag{3.17}$$

b. Statically indeterminate structure

In this case,

$$\{D\} = \begin{Bmatrix} \{D_A\} \\ \hline \{D_x\} \end{Bmatrix} \text{ and } \{P\} = \begin{Bmatrix} \{P_A\} \\ \hline \{P_x\} \end{Bmatrix}$$

Hence,

$$\{D\} = [F]\{P\}$$

$$\left\{ \begin{array}{c} \{D_A\} \\ \hline \{D_X\} \end{array} \right\} = \left[\begin{array}{c|c} [F_{AA}] & [F_{AX}] \\ \hline [F_{XA}] & [F_{XX}] \end{array} \right] \left\{ \begin{array}{c} \{P_A\} \\ \hline \{P_A\} \end{array} \right\} \tag{3.18}$$

Thus,

$$\{D_A\} = [F_{AA}]\{P_A\} + [F_{AX}]\{P_X\} \tag{3.19a}$$

$$\{D_X\} = [F_{XA}]\{P_A\} + [F_{XX}]\{P_X\} \tag{3.19b}$$

For a statically indeterminate structure, the structure flexibility matrix is

$$[F] = \left[\begin{array}{c|c} [F_{AA}] & [F_{AX}] \\ \hline [F_{XA}] & [F_{XX}] \end{array} \right] \tag{3.20}$$

In the above equation, $[F]$ is subdivided into four submatrices to consider the effects of $\{P_A\}$ and $\{P_X\}$.

3.5 Compatibility Conditions

The compatibility condition is the condition on the continuity of displacements in the loaded structure. The equilibrium equations alone are not sufficient for the analysis of statically indeterminate structures. In the analysis of these types of structures, the compatibility condition is introduced by using $\{D_X\}$, the redundant displacement vector. $\{D_X\}$ is the known displacement vector corresponding to the unknown redundant vector $\{P_X\}$.

The compatibility conditions used in the flexibility method of analysis are that in a primary structure, the sum of displacements at the redundant locations due to loads and redundants must be equal to $\{D_X\}$. If the $\{D_{XA}\}$ and $\{D_{XX}\}$ are the displacements at the redundant coordinates due to load vector $\{P_A\}$ and redundants $\{P_X\}$, then the compatibility conditions are

$$\{D_{XA}\} + \{D_{XX}\} = \{D_X\}$$

where $\{D_{XA}\} = [F_{XA}]\{P_A\}$ and $\{D_{XX}\} = [F_{XX}]\{P_X\}$
The above equation is same as Eq. 3.19b.

3.6 Structure Flexibility Matrix

The law of conservation of energy is used to derive the expression of $[F]$. The law states that the work done by the loads W is equal to the strain energy U stored in the structure. The work done by the loads W is

$$W = \frac{1}{2}\{P\}^T \{D\}$$

Strain energy U is the work done by the internal forces $\{p\}$ over the internal displacement $\{d\}$. Therefore,

$$U = \frac{1}{2}\{p\}^T \{d\}$$

Hence,

$$W = U \Rightarrow \frac{1}{2}\{P\}^T \{D\} = \frac{1}{2}\{p\}^T \{d\} \qquad (a)$$

But $\{p\} = [T_F] \{P\}$ (Eq. 3.5)

$$\Rightarrow \quad \{p\}^T = \{P\}^T [T_F]^T \qquad (b)$$

Also $\{d\} = [f] \{p\} = [f] [T_F] \{P\}$ (Using Eqs. 3.14 and 3.5) $\qquad$ (c)
$\quad$ Substituting Eqs. b and c in Eq. a gives

$$\{P\}^T \{D\} = \{P\}^T [T_F]^T [f][T_F]\{P\}$$

$$\Rightarrow \quad \{D\} = [T_F]^T [f][T_F]\{P\}$$

But $\{D\} = [F] \{P\}$ (Eq. 3.15)
$\quad$ Thus, the flexibility matrix of the structure $[F]$ is

$$[F] = [T_F]^T [f][T_F] \qquad (3.21)$$

3.6.1 Flexibility Matrix of a Statically Determinate Structure

For a statically determinate structure,
$\quad [T_F] = [T_{FA}]$ and $[F] = [F_{AA}]$ (Eqs. 3.7 and 3.17)
$\quad$ Hence, from Eq. 3.21,

$$[F] = [F_{AA}] = [T_{FA}]^T [f][T_{FA}] \qquad (3.22)$$

3.6.2 Flexibility Matrix of a Statically Indeterminate Structure

In this case,

$$[T_F] = \left[\begin{array}{c:c} [T_{FX}] & [T_{FX}] \end{array} \right] \qquad \text{(Eq. 3.9)}$$

Hence,

$$[F] = [T_F]^T [f][T_F]$$

$$= \left[\begin{array}{c} [T_{FA}]^T \\ \hline [T_{FX}]^T \end{array} \right] [f] \left[\begin{array}{c|c} [T_{FA}] & [T_{FX}] \end{array} \right]$$

$$= \left[\begin{array}{c|c} [T_{FA}]^T[f][T_{FA}] & [T_{FA}]^T[f][T_{FX}] \\ \hline [T_{FX}]^T[f][T_{FA}] & [T_{FX}]^T[f][T_{FX}] \end{array} \right] \tag{3.23}$$

But as per Eq. 3.20,

$$[F] = \left[\begin{array}{c|c} [F_{AA}] & [F_{AX}] \\ \hline [F_{XA}] & [F_{XX}] \end{array} \right]$$

Comparing Eqs. 3.20 and 3.23 gives

$$[F_{AA}] = [T_{FA}]^T[f][T_{FA}] \tag{3.24a}$$

$$[F_{AX}] = [T_{FA}]^T[f][T_{FX}] \tag{3.24b}$$

$$[F_{XA}] = [T_{FX}]^T[f][T_{FA}] \tag{3.24c}$$

$$[F_{XX}] = [T_{FX}]^T[f][T_{FX}] \tag{3.24d}$$

3.6.3 Relation between $[F_{AX}]$ and $[F_{XA}]$

From Eq. 3.24b,

$$[F_{AX}] = [T_{FA}]^T[f][T_{FX}] = [T_{FA}]^T \left([f][T_{FX}] \right)$$

Taking the transpose of the matrix,

$$[F_{AX}]^T = \left([f][T_{FX}] \right)^T \left([T_{FA}]^T \right)^T$$

$$= [T_{FX}]^T[f]^T[T_{FA}]$$

But $[f]^T = [f]$ ($\because [f]$ is a diagonal matrix)
   Hence,

$$[F_{AX}]^T = [T_{FX}]^T[f][T_{FA}] = [F_{XA}]$$

Thus,

$$[F_{AX}]^T = [F_{XA}] \tag{3.25}$$

3.7 Transformations Used in Flexibility Method

Figure 3.9 shows the various transformations used in the flexibility method.
From Figure 3.9,

$$\{p\} = [T_F]\{P\}$$

$$\{d\} = [f]\{p\} = [f][T_F]\{P\}$$

$$\{D\} = [a]\{d\} = [T_F]^T\{d\}$$

Note: Using the principle of contragradient discussed in Section 4.4, it can be shown that if $\{p\} = [T_F]\{P\}$, then $\{D\} = [a]\{d\}$, where $[a] = [T_F]^T$.
    Hence,

$$\{D\} = [T_F]^T\{d\} = [T_F]^T[f][T_F]\{P\} = [F]\{P\}$$

$$\therefore \{F\} = [T_F]^T[f][T_F]$$

(This is an alternate way for deriving the expression for finding [F].)

3.8 Analysis of Statically Determinate Structure

The equations of equilibrium alone are sufficient to determine the element forces in a statically determinate structure and they are determined using the following equation:

$$\{p\} = [T_{FA}]\{P_A\}$$

The nodal displacements $\{D_A\}$ are calculated by the equation

$$\{D_A\} = [F_{AA}]\{P_A\}$$

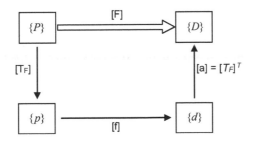

FIGURE 3.9
Transformation matrices in the flexibility method.

The steps required to analyze a statically determinate structure using the flexibility method are listed below.

   i. Identify the nodes and elements and draw the discretized structure.
   ii. Identify the global coordinates (action coordinates) and define the load vector $\{P_A\}$
   iii. Identify the local coordinates. Define the element force vector $\{p^i\}$ and combined element force vector $\{p\}$.
   iv. Determine the force transformation matrix $[T_{FA}]$.
   v. Find the element forces $\{p\}$ using the equation $\{p\} = [T_{FA}]\{P_A\}$.
   vi. Determine the individual element flexibility matrices $[f^1]$, $[f^2]$, ... $[f^n]$ and develop the unassembled flexibility matrix $[f]$.

$$[f] = \begin{bmatrix} [f^1] & 0 & 0 & \cdots & 0 \\ 0 & [f^2] & 0 & \cdots & 0 \\ \cdots & \cdots & \cdots & \cdots & \cdots \\ 0 & 0 & 0 & \cdots & [f^n] \end{bmatrix}$$

   vii. Determine the flexibility matrix of the structure $[F_{AA}]$.

$$[F_{AA}] = [T_{FA}]^T [f][T_{FA}]$$

   viii. Calculate the nodal displacement $\{D_A\}$ using the equation $\{D_A\} = [F_{AA}]\{P_A\}$.

3.8.1 Structures Subjected to Element Loads

As discussed in Chapter 1, the loads acting on the elements are converted into equivalent nodal loads $\{P_e\}$ ($\{P_e\} = -\{P_F\}$). After finding $\{P_e\}$, the combined nodal force vector $\{P_C\}$ is calculated ($\{P_C\} = \{P\} + \{P_e\}$). In the case of statically determinate structures, $\{P_{CA}\}$ (i.e., combined loads along the action coordinates) are used in the place of $\{P_A\}$ in the above steps.

Using the principle of superposition, the element forces $\{p\}$ are obtained as

$$\{p\} = \{p_F\} + \{p_C\} \tag{3.26}$$

where $\{p_F\}$ is the element force due to fixed end actions and $\{p_C\}$ is the element forces due to combined nodal loads ($\{p_C\} = [T_{FA}]\{P_{CA}\}$).

The equivalent nodal loads are determined such that the nodal displacements due to combined loads $\{D_C\}$ will be same as the nodal displacements due to actual loads $\{D\}$. Thus,

$$\{D_C\} = \{D\}$$

Hence,

$$\{D_C\} = [F]\{P_C\}$$

$$\Rightarrow \quad \{D\} = [F]\{P_C\}$$

In the case of statically determinate structure,

$$\{D_A\} = [F_{AA}]\{P_C\} \quad (\because [F] = [F_{AA}] \text{ and } \{D\} = \{D_A\})$$

Example 3.2

Find the deflection and rotation at the free end of the beam shown in Figure 3.10a. Also find the forces developed in the beam.

Solution

Step 1: Develop the discretized structure
   The nodes and element are shown in Figure 3.10b.
Step 2: Identify the global coordinates and define load vector
   Since it is necessary to find vertical deflection and rotation at node 2, two global coordinates are identified at node 2 (Figure 3.10c). The load vector is

$$\{P_A\} = \left\{ \begin{array}{c} P_1 \\ P_2 \end{array} \right\} = \left\{ \begin{array}{c} -P \\ 0 \end{array} \right\}$$

Step 3: Identify the local coordinates
   The structure is modeled using a beam element. The combined local coordinates are shown in Figure 3.10d.

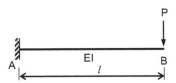

FIGURE 3.10a
Example 3.2.

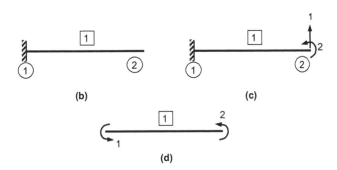

FIGURE 3.10b–d
Discretized structure and coordinates. (b) Discretized structure. (c) Global coordinates. (d) Combined local coordinates.

Step 4: Develop the force transformation matrix $[T_{FA}]$

Unit loads are applied at the two global coordinates and the end moments developed due to these loads are shown in Figure 3.10e. Comparing with the combined local coordinates shown in Figure 3.10d, the force transformation matrix is obtained as

$$[T_{FA}] = \begin{matrix} \overset{P_1=1}{} & \overset{P_2=1}{} \\ \begin{bmatrix} -l & -1 \\ 0 & 1 \end{bmatrix} \end{matrix}$$

Step 5: Develop the element flexibility matrix $[f^i]$ and unassembled flexibility matrix $[f]$

- Element flexibility matrix

$$[f^1] = \frac{l}{6EI} \begin{bmatrix} 2 & -1 \\ -1 & 2 \end{bmatrix}$$

- Unassembled flexibility matrix $[f]$
  Since there is only one element,

$$[f] = [f^1] = \frac{l}{6EI} \begin{bmatrix} 2 & -1 \\ -1 & 2 \end{bmatrix}$$

Step 6: Develop flexibility matrix $[F_{AA}]$

$$[F_{AA}] = [T_{FA}]^T [f][T_{FA}]$$

$$[F_{AA}] = \begin{bmatrix} -l & 0 \\ -1 & 1 \end{bmatrix} \frac{l}{6EI} \begin{bmatrix} 2 & -1 \\ -1 & 2 \end{bmatrix} \begin{bmatrix} -l & -1 \\ 0 & 1 \end{bmatrix}$$

$$= \frac{l}{6EI} \begin{bmatrix} 2l^2 & 3l \\ 3l & 6 \end{bmatrix}$$

Step 7: Determine the nodal displacement $\{D_A\}$

$$\{D_A\} = [F_{AA}]\{P_A\}$$

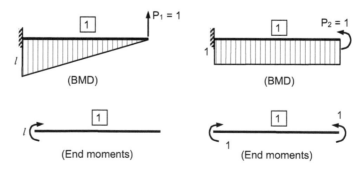

FIGURE 3.10e
Elements of $[T_{FA}]$.

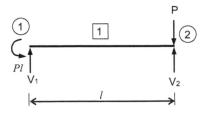

FIGURE 3.10f
End actions in the element.

$$\{D_A\} = \left\{ \begin{array}{c} D_1 \\ D_2 \end{array} \right\} = \frac{l}{6EI} \left[\begin{array}{cc} 2l^2 & 3l \\ 3l & 6 \end{array} \right] \left\{ \begin{array}{c} -P \\ 0 \end{array} \right\} = \left\{ \begin{array}{c} -Pl^3/3EI \\ -Pl^2/2EI \end{array} \right\}$$

Hence, the vertical deflection and rotation at the tip of the beam are $Pl^3/3EI$ (downward) and $Pl^2/2EI$ (clockwise).

Step 8: Determine the element forces $\{p\}$

$$\{p\} = [T_{FA}]\{P_A\}$$

$$\{p\} = \left\{ \begin{array}{c} p_1 \\ p_2 \end{array} \right\} = \left[\begin{array}{cc} -l & -1 \\ 0 & 1 \end{array} \right] \left\{ \begin{array}{c} -P \\ 0 \end{array} \right\} = \left\{ \begin{array}{c} Pl \\ 0 \end{array} \right\}$$

Once $\{p\}$ are calculated, the end actions (Figure 3.10f) are determined using equations of equilibrium.

From Figure 3.10f,

$$\sum M_1 = 0 \Rightarrow Pl + lV_2 - lP = 0 \Rightarrow V_2 = 0$$

$$\sum F_Y = 0 \Rightarrow V_1 = P$$

After finding end actions, the shear force diagram (SFD) and BMD of the structures can be drawn.

Example 3.3

Determine the deflection at the free end of the cantilever beam shown in Figure 3.11a.

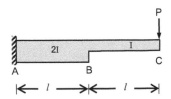

FIGURE 3.11a
Example 3.3.

Solution

Step 1: Develop the discretized structure
  Since there is a change in cross-section at B, it is necessary to introduce a node at B. The beam is modeled using two beam elements (Figure 3.11b).
Step 2: Identify the global coordinates and define load vector
  The global coordinate 1 shown in Figure 3.11c is sufficient to define the load and vertical deflection at joint C. The load vector is

$$\{P_A\} = \{P_1\} = \{-P\}$$

Step 3: Identify the local coordinates
  The combined local coordinates for the two elements are shown in Figure 3.11d.
Step 4: Develop the force transformation matrix $[T_{FA}]$
  To determine the elements of $[T_{FA}]$, a unit load is applied at the global coordinate and the element forces (end moments) developed due to this load are shown in Figure 3.11e.
  By comparing the moments developed in the elements (Figure 3.11e) with the combined local coordinates (Figure 3.11d), the elements of $[T_{FA}]$ are

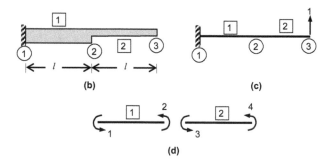

(b)   (c)

(d)

FIGURE 3.11b–d
Discretized structure and coordinates. (b) Discretized structure. (c) Global coordinates. (d) Combined local coordinates.

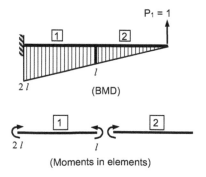

FIGURE 3.11e
Elements of $[T_{FA}]$.

$$[T_{FA}] = \begin{bmatrix} -2l \\ l \\ -l \\ 0 \end{bmatrix}$$

Step 5: Develop the element flexibility matrix $[f^i]$ and unassembled flexibility matrix $[f]$

- Element flexibility matrix
  The flexibility matrix of the beam element is

$$[f^i] = \frac{l_i}{6(EI)_i} \begin{bmatrix} 2 & -1 \\ -1 & 2 \end{bmatrix}$$

Thus,

$$[f^1] = \frac{l}{6(2EI)} \begin{bmatrix} 2 & -1 \\ -1 & 2 \end{bmatrix} = \frac{l}{6EI} \begin{bmatrix} 1 & -0.5 \\ -0.5 & 1 \end{bmatrix}$$

$$[f^2] = \frac{l}{6EI} \begin{bmatrix} 2 & -1 \\ -1 & 2 \end{bmatrix}$$

- Unassembled flexibility matrix

$$[f] = \begin{bmatrix} [f^1] & 0 \\ 0 & [f^2] \end{bmatrix}$$

$$= \frac{l}{6EI} \begin{bmatrix} 1 & -0.5 & 0 & 0 \\ -0.5 & 1 & 0 & 0 \\ 0 & 0 & 2 & -1 \\ 0 & 0 & -1 & 2 \end{bmatrix}$$

Step 6: Develop the flexibility matrix $[F_{AA}]$

$$[F_{AA}] = [T_{FA}]^T [f][T_{FA}]$$

$$= \begin{bmatrix} -2l & l & -l & 0 \end{bmatrix} \frac{l}{6EI} \begin{bmatrix} 1 & -0.5 & 0 & 0 \\ -0.5 & 1 & 0 & 0 \\ 0 & 0 & 2 & -1 \\ 0 & 0 & -1 & 2 \end{bmatrix} \begin{bmatrix} -2l \\ l \\ -l \\ 0 \end{bmatrix}$$

$$= \frac{3}{2} \frac{l^3}{EI}$$

Step 7: Determine the nodal displacements $\{D_A\}$

$$\{D_A\} = [F_{AA}]\{P_A\}$$

$$\Rightarrow \{D_A\} = \{D_1\} = \frac{3}{2}\frac{l^3}{EI} \times (-P) = -\frac{3}{2}\frac{Pl^3}{EI}$$

Hence, the displacement at the tip of the beam is $\dfrac{3}{2}\dfrac{Pl^3}{EI}(\downarrow)$

Example 3.4

Determine the deflection at the free end of the beam shown in Figure 3.12a.

Solution

Step 1: Develop the discretized structure
   The load is assumed to act at a node. Hence, three nodes are necessary (at A, B and C). The discretized structure is shown in Figure 3.12b.
Step 2: Identify the global coordinates and define load vector
   In order to define the applied vertical load at node 2 and vertical deflection at node 3, two global coordinates are used (Figure 3.12c). The load vector $\{P_A\}$ is

$$\{P_A\} = \left\{ \begin{array}{c} P_1 \\ P_2 \end{array} \right\} = \left\{ \begin{array}{c} -P \\ 0 \end{array} \right\}$$

Step 3: Identify the local coordinates
   The combined local coordinates for the two elements are shown in Figure 3.12d.
Step 4: Develop the force transformation matrix $[T_{FA}]$
   The end moments developed in the elements due to unit loads acting at the global coordinates are shown in Figure 3.12e.
   From Figures 3.12d and e,

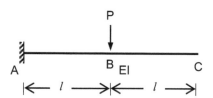

FIGURE 3.12a
Example 3.4.

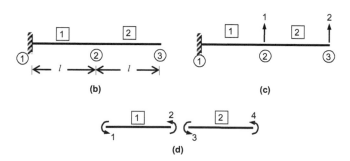

FIGURE 3.12b–d
Discretized structure and coordinates. (b) Discretized structure. (c) Global coordinates. (d) Combined local coordinates.

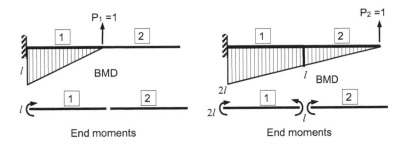

FIGURE 3.12e
Elements of $[T_{FA}]$.

$$[T_{FA}] = \begin{bmatrix} \overset{P_1=1}{-l} & \overset{P_2=1}{-2l} \\ 0 & l \\ 0 & -l \\ 0 & 0 \end{bmatrix}$$

Step 5: Develop the element flexibility matrix $[f^i]$ and unassembled flexibility matrix $[f]$.

- Element flexibility matrix
  The flexibility matrix of the beam element is

$$[f^1] = [f^2] = \frac{l}{6EI} \begin{bmatrix} 2 & -1 \\ -1 & 2 \end{bmatrix}$$

- Unassembled flexibility matrix

$$[f] = \begin{bmatrix} [f_1] & 0 \\ 0 & [f_2] \end{bmatrix} = \frac{l}{6EI} \begin{bmatrix} 2 & -1 & 0 & 0 \\ -1 & 2 & 0 & 0 \\ 0 & 0 & 2 & -1 \\ 0 & 0 & -1 & 2 \end{bmatrix}$$

Step 6: Develop the flexibility matrix $[F_{AA}]$

$$[F_{AA}] = [T_{FA}]^T [f][T_{FA}] = \frac{l^3}{6EI} \begin{bmatrix} 2 & 5 \\ 5 & 16 \end{bmatrix}$$

Step 7: Determine the nodal displacements $\{D_A\}$

$$\{D_A\} = [F_{AA}]\{P_A\}$$

$$\{D_A\} = \begin{Bmatrix} D_1 \\ D_2 \end{Bmatrix} = \frac{l^3}{6EI} \begin{bmatrix} 2 & 5 \\ 5 & 16 \end{bmatrix} \begin{Bmatrix} -P \\ 0 \end{Bmatrix} = \begin{Bmatrix} -Pl^3/3EI \\ -5Pl^3/6EI \end{Bmatrix}$$

Hence, the vertical deflection at the tip of the beam is $\dfrac{5Pl^3}{6EI}(\downarrow)$

Example 3.5

Find the vertical deflection at joint A of the structure shown in Figure 3.13a.

Solution

Step 1: Develop the discretized structure
   The nodes and elements are shown in Figure 3.13b.
Step 2: Identify the global coordinates and define load vector $\{P_A\}$
   The global coordinates are shown in Figure 3.13c and the load vector is

$$\{P_A\} = \left\{ \begin{array}{c} P_1 \\ P_2 \end{array} \right\} = \left\{ \begin{array}{c} -5\,\text{kN} \\ 2\,\text{kNm} \end{array} \right\}$$

Step 3: Identify the local coordinates
   The combined local coordinates of the elements are shown in Figure 3.13d.
Step 4: Develop the force transformation matrix $[T_{FA}]$
   The end moments (in kNm) developed in elements due to the application of unit loads acting at the global coordinates are shown in Figure 3.13e.
   From Figures 3.13d and e,

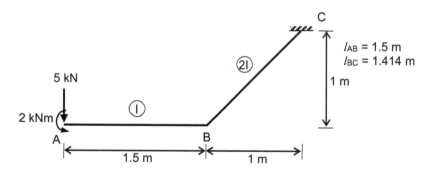

FIGURE 3.13a
Example 3.5.

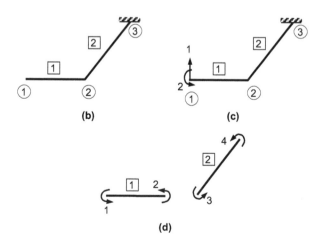

FIGURE 3.13b–d
Discretized structure and coordinates. (b) Discretized structure. (c) Global coordinates. (d) Combined local coordinates.

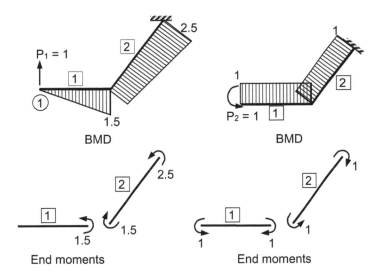

FIGURE 3.13e
Elements of $[T_{FA}]$.

$$[T_{FA}] = \begin{array}{cc} \scriptstyle P_1=1 & \scriptstyle P_2=1 \\ \begin{bmatrix} 0 & 1 \\ 1.5 & -1 \\ -1.5 & 1 \\ 2.5 & -1 \end{bmatrix} \end{array}$$

Step 5: Develop the element flexibility matrix $[f^i]$ and unassembled flexibility matrix $[f]$.

- Element flexibility matrix $[f^i]$

$$[f^1] = \frac{1.5}{6EI}\begin{bmatrix} 2 & -1 \\ -1 & 2 \end{bmatrix} = \frac{1}{6EI}\begin{bmatrix} 3 & -1.5 \\ -1.5 & 3 \end{bmatrix}$$

$$[f^2] = \frac{1.414}{6(2EI)}\begin{bmatrix} 2 & -1 \\ -1 & 2 \end{bmatrix} = \frac{1}{6EI}\begin{bmatrix} 1.414 & -0.707 \\ -0.707 & 1.414 \end{bmatrix}$$

- Unassembled flexibility matrix

$$[f] = \begin{bmatrix} [f^1] & 0 \\ 0 & [f^2] \end{bmatrix} = \frac{1}{6EI}\begin{bmatrix} 3 & -1.5 & 0 & 0 \\ -1.5 & 3 & 0 & 0 \\ 0 & 0 & 1.414 & -0.707 \\ 0 & 0 & -0.707 & 1.414 \end{bmatrix}$$

Step 6: Develop the flexibility matrix $[F_{AA}]$

$$[F_{AA}] = [T_{FA}]^T[f][T_{FA}] = \frac{1}{EI}\begin{bmatrix} 4.012 & -2.539 \\ -2.539 & 2.207 \end{bmatrix}$$

Step 7: Determine the nodal displacements $\{D_A\}$

$$\{D_A\} = [F_{AA}]\{P_A\}$$

$$\{D_A\} = \left\{ \begin{array}{c} D_1 \\ D_2 \end{array} \right\} = \frac{1}{EI} \left\{ \begin{array}{c} -25.138 \\ 17.109 \end{array} \right\}$$

Hence, the vertical deflection at joint A is $\dfrac{25.138}{EI} (\downarrow)$

Example 3.6

Find the horizontal displacement at joint D of the frame shown in Figure 3.14a.

Solution

Step 1: Develop the discretized structure
   The discretized structure is shown in Figure 3.14b.
Step 2: Identify the global coordinates and define load vector
   The global coordinates are shown in Figure 3.14c and the load vector is

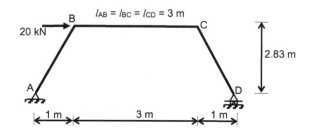

FIGURE 3.14a
Example 3.6.

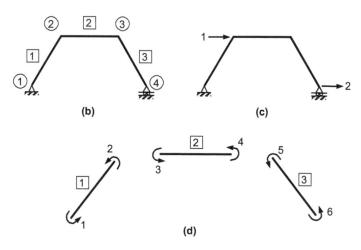

FIGURE 3.14b–d
Discretized structure and coordinates. (b) Discretized structure. (c) Global coordinates. (d) Combined local coordinates.

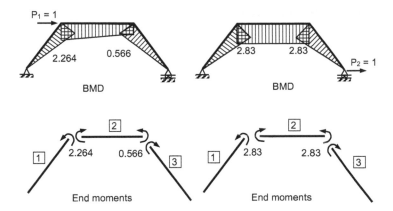

FIGURE 3.14e
Elements of $[T_{FA}]$.

$$\{P_A\} = \left\{ \begin{array}{c} P_1 \\ P_2 \end{array} \right\} = \left\{ \begin{array}{c} 20 \\ 0 \end{array} \right\} kN$$

Step 3: Identify the local coordinates

The combined local coordinates are shown in Figure 3.14d.

Step 4: Develop the force transformation matrix $[T_{FA}]$

The end moments (in kNm) developed in elements due to unit loads applied at the global coordinates are shown in Figure 3.14e.

From Figures 3.14d and e,

$$[T_{FA}] = \begin{bmatrix} P_1 = 1 & P_2 = 1 \\ 0 & 0 \\ 2.264 & 2.83 \\ -2.264 & -2.83 \\ 0.566 & 2.83 \\ -0.566 & -2.83 \\ 0 & 0 \end{bmatrix}$$

Step 5: Develop the element flexibility matrix and unassembled flexibility matrix.

• Element flexibility matrix $[f^i]$

$$[f^1] = [f^2] = [f^3] = \frac{3}{6EI}\begin{bmatrix} 2 & -1 \\ -1 & 2 \end{bmatrix} = \frac{1}{6EI}\begin{bmatrix} 6 & -3 \\ -3 & 6 \end{bmatrix}$$

• Unassembled flexibility matrix $[f]$

$$[f] = \begin{bmatrix} [f^1] & 0 & 0 \\ 0 & [f^2] & 0 \\ 0 & 0 & [f^3] \end{bmatrix} = \frac{1}{6EI}\begin{bmatrix} 6 & 3 & 0 & 0 & 0 & 0 \\ -3 & 6 & 0 & 0 & 0 & 0 \\ 0 & 0 & 6 & -3 & 0 & 0 \\ 0 & 0 & -3 & 6 & 0 & 0 \\ 0 & 0 & 0 & 0 & 6 & -3 \\ 0 & 0 & 0 & 0 & -3 & 6 \end{bmatrix}$$

Step 6: Develop the flexibility matrix $[F_{AA}]$

$$[F_{AA}] = [T_{FA}]^T [f][T_{FA}]$$

$$= \frac{1}{EI}\begin{bmatrix} 12.174 & 20.022 \\ 20.022 & 40.045 \end{bmatrix}$$

Step 7: Determine the nodal displacements

$$\{D_A\} = [F_{AA}]\{P_A\}$$

$$\{D_A\} = \begin{Bmatrix} D_1 \\ D_2 \end{Bmatrix} = \frac{1}{EI}\begin{Bmatrix} 243.471 \\ 400.445 \end{Bmatrix}$$

The horizontal deflection at joint D is $\dfrac{400.445}{EI}(\rightarrow)$

Example 3.7

Find the force developed in the truss shown in Figure 3.15a. Also find the displacements corresponding to the applied loads. The ratio of length (l_i) to cross-section area (A_i) of all elements is unity.

Solution

Step 1: Develop the discretized structure
   The discretized structure indicating the nodes and elements are shown in Figure 3.15b.
Step 2: Identify the global coordinates and define load vector
   The global coordinates are shown in Figure 3.15c and the load vector is

$$\{P_A\} = \begin{Bmatrix} P_1 \\ P_2 \end{Bmatrix} = \begin{Bmatrix} -2P \\ P \end{Bmatrix}$$

Step 3: Identify the local coordinates
   The combined local coordinates are shown in Figure 3.15d.

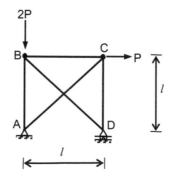

FIGURE 3.15a
Example 3.7.

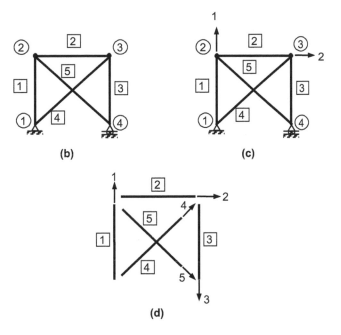

FIGURE 3.15b–d
Discretized structure and coordinates. (b) Discretized structure. (c) Global coordinates. (d) Combined local coordinates.

Step 4: Develop the force transformation matrix $[T_{FA}]$

The forces developed in the elements due to unit loads applied at the global coordinates are shown in Figure 3.15e.

From the figure,

$$[T_{FA}] = \begin{matrix} & {\scriptstyle P_1=1} & {\scriptstyle P_2=1} \\ \left[\begin{matrix} 1 & 0 \\ 0 & 0 \\ 0 & -1 \\ 0 & \sqrt{2} \\ 0 & 0 \end{matrix} \right] \end{matrix}$$

Step 5: Develop the element flexibility matrix and unassembled flexibility matrix.

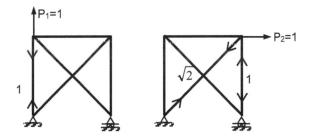

FIGURE 3.15e
Elements of $[T_{FA}]$.

- Element flexibility matrix $[f^i]$
  For a truss element, $[f^i] = [l_i/A_iE_i]$

$$\left[f^1\right] = \left[f^2\right] = \left[f^3\right] = \left[f^4\right] = \left[f^5\right] = \left[1/E\right]$$

- Unassembled flexibility matrix $[f]$

$$[f] = \begin{bmatrix} \left[f^1\right] & 0 & 0 & 0 & 0 \\ 0 & \left[f^2\right] & 0 & 0 & 0 \\ 0 & 0 & \left[f^3\right] & 0 & 0 \\ 0 & 0 & 0 & \left[f^4\right] & 0 \\ 0 & 0 & 0 & 0 & \left[f^5\right] \end{bmatrix} = \frac{1}{E}\begin{bmatrix} 1 & 0 & 0 & 0 & 0 \\ 0 & 1 & 0 & 0 & 0 \\ 0 & 0 & 1 & 0 & 0 \\ 0 & 0 & 0 & 1 & 0 \\ 0 & 0 & 0 & 0 & 1 \end{bmatrix}$$

Step 6: Develop the flexibility matrix $[F_{AA}]$

$$[F_{AA}] = [T_{FA}]^T[f][T_{FA}] = \frac{1}{E}\begin{bmatrix} 1 & 0 \\ 0 & 3 \end{bmatrix}$$

Step 7: Determine the nodal displacements $\{D_A\}$

$$\{D_A\} = [F_{AA}]\{P_A\}$$

$$\{D_A\} = \left\{ \begin{array}{c} D_1 \\ D_2 \end{array} \right\} = \left\{ \begin{array}{c} -2 \\ 3 \end{array} \right\}\frac{P}{E}$$

Step 8: Determine the element forces $\{p\}$

$$\{p\} = [T_{FA}]\{P_A\}$$

$$\{p\} = \left\{ \begin{array}{c} p_1 \\ p_2 \\ p_3 \\ p_4 \end{array} \right\} = \left\{ \begin{array}{c} -2 \\ 0 \\ -1 \\ \sqrt{2} \\ 0 \end{array} \right\}P$$

Example 3.8

In the previous example, determine all the nodal displacements of the truss.

Solution

Step 1: Develop the discretized structure
  The nodes and elements are shown in Figure 3.15b.
Step 2: Identify the global coordinates and define load vector
  In this case, it is necessary to find all nodal displacements. Hence, all the active coordinates are treated as the global coordinates (Figure 3.16a) and the load vector is

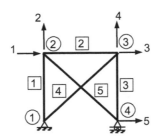

FIGURE 3.16a
Global coordinates.

$$\{P_A\} = \left\{ \begin{array}{c} P_1 \\ P_2 \\ P_3 \\ P_4 \\ P_5 \end{array} \right\} = \left\{ \begin{array}{c} 0 \\ -2P \\ P \\ 0 \\ 0 \end{array} \right\}$$

Step 3: Identify the local coordinates
  The combined local coordinates are shown in Figure 3.15d.
Step 4: Develop the force transformation matrix $[T_{FA}]$
  The forces developed in the truss due to unit loads applied at different global coordinates are shown in Figure 3.16b.
  The force transformation matrix is

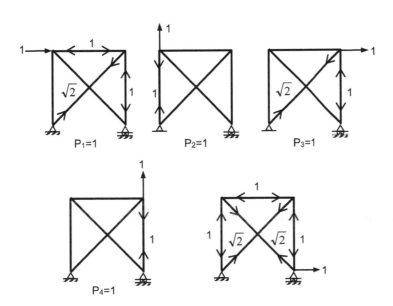

FIGURE 3.16b
Elements of $[T_{FA}]$.

$$P_1=1 \quad P_2=1 \quad P_3=1 \quad P_4=1 \quad P_5=1$$

$$[T_{FA}] = \begin{bmatrix} 0 & 1 & 0 & 0 & -1 \\ -1 & 0 & 0 & 0 & -1 \\ -1 & 0 & -1 & 1 & -1 \\ \sqrt{2} & 0 & \sqrt{2} & 0 & \sqrt{2} \\ 0 & 0 & 0 & 0 & \sqrt{2} \end{bmatrix}$$

Step 5: Develop $[f^i]$ and $[f]$

$$\left[f^i \right] = \left[1/E \right]$$

$$[f] = \frac{1}{E} \begin{bmatrix} 1 & 0 & 0 & 0 & 0 \\ 0 & 1 & 0 & 0 & 0 \\ 0 & 0 & 1 & 0 & 0 \\ 0 & 0 & 0 & 1 & 0 \\ 0 & 0 & 0 & 0 & 1 \end{bmatrix}$$

Step 6: Develop the flexibility matrix $[F_{AA}]$

$$[F_{AA}] = [T_{FA}]^T [f][T_{FA}]$$

$$= \frac{1}{E} \begin{bmatrix} 4 & 0 & 3 & -1 & 4 \\ 0 & 1 & 0 & 0 & -1 \\ 3 & 0 & 3 & -1 & 3 \\ -1 & 0 & -1 & 1 & -1 \\ 4 & -1 & 3 & -1 & 7 \end{bmatrix}$$

Step 7: Determine the nodal displacements $\{D_A\}$

$$\{D_A\} = [F_{AA}]\{P_A\}$$

$$\{D_A\} = \begin{Bmatrix} D_1 \\ D_2 \\ D_3 \\ D_4 \\ D_5 \end{Bmatrix} = \frac{P}{E} \begin{Bmatrix} 3 \\ -2 \\ 3 \\ -1 \\ 5 \end{Bmatrix}$$

Example 3.9

Determine the deflection at joint C of the beam shown in Figure 3.17a. Also draw the SFD and BMD.

Solution

Step 1: Develop the discretized structure.

The load P is treated as an element load. Hence, the structure is modeled using a single beam element (Figure 3.17b).

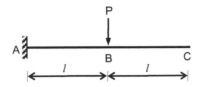

FIGURE 3.17a
Example 3.9.

Step 2: Determine the combined nodal loads

Since P is treated as an element load, it is necessary to replace P by its equivalent nodal loads. The restrained structure, fixed end actions, and equivalent nodal loads are shown in Figures 3.17c and d. Since there are no other nodal loads, the equivalent nodal loads shown in Figure 3.17d will be the combined nodal loads.

Step 3: Identify the global coordinates and define the load vector

To get a response in a structure, it is necessary that the loads are applied along the active coordinates. If the loads are applied at the restrained coordinates (i.e., at the supports), then they will not cause any response in the structure. These loads will be directly transferred to the supports. Hence, to predict the response of the structure, it is sufficient to consider the loads acting only at the active coordinates. In this case, the loads at node 2 are along the active coordinate. Hence, the two action coordinates are considered (Figure 3.17e) and the load vector is

$$\{P_{CA}\} = \left\{ \begin{array}{c} P_{C1} \\ P_{C2} \end{array} \right\} = \left\{ \begin{array}{c} -P/2 \\ Pl/4 \end{array} \right\}$$

Step 4: Identify the local coordinates and find $\{p_F\}$

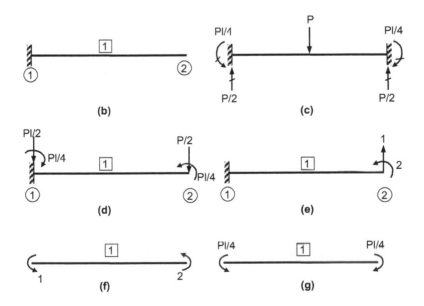

FIGURE 3.17b–g
Discretized structure, fixed end actions, and coordinates. (b) Discretized structure. (c) Restrained structure and fixed end actions. (d) Equivalent nodal loads. (e) Global Coordinates. (f) Combined local coordinates. (g) End moment in elements due to fixed end actions.

The combined local coordinates are shown in Figure 3.17f.

- Element forces due to fixed end actions $\{p_F\}$

  The end moments developed in the element due to fixed end action (Figure 3.17c) are shown in Figure 3.17g. Comparing with the direction of moments shown in Figure 3.17f gives,

$$\{p_F\} = \left\{ \begin{array}{c} Pl/4 \\ -Pl/4 \end{array} \right\}$$

Step 5: Develop the force transformation matrix $[T_{FA}]$

The end moments developed in the elements due to application of unit loads along the action coordinates are shown in Figure 3.17h.

Hence,

$$[T_{FA}] = \begin{bmatrix} \overset{P_{c1}=1}{-2l} & \overset{P_{c2}=1}{-1} \\ 0 & 1 \end{bmatrix}$$

Step 6: Develop $[f^1]$ and $[f]$

- Element flexibility matrix

$$[f^1] = \frac{(2l)}{6EI}\begin{bmatrix} 2 & -1 \\ -1 & 2 \end{bmatrix} = \frac{l}{3EI}\begin{bmatrix} 2 & -1 \\ -1 & 2 \end{bmatrix}$$

- Unassembled flexibility matrix

$$[f] = [f^1] = \frac{l}{3EI}\begin{bmatrix} 2 & -1 \\ -1 & 2 \end{bmatrix}$$

Step 7: Develop the flexibility matrix $[F_{AA}]$

$$[F_{AA}] = [T_{FA}]^T [f][T_{FA}]$$

$$= \frac{l}{3EI}\begin{bmatrix} 8l^2 & 6l \\ 6l & 6 \end{bmatrix}$$

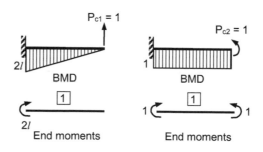

FIGURE 3.17h
Elements of $[T_{FA}]$.

Step 8: Determine the nodal displacements $\{D_A\}$

$$\{D_A\} = [F_{AA}]\{P_{CA}\}$$

$$= \frac{l}{3EI} \begin{bmatrix} 8l^2 & 6l \\ 6l & 6 \end{bmatrix} \begin{Bmatrix} -P/2 \\ Pl/4 \end{Bmatrix} = \begin{Bmatrix} -5/6 & Pl^3/EI \\ -3/6 & Pl^2/EI \end{Bmatrix}$$

Hence, the deflection at joint C is $\dfrac{5}{6}\dfrac{Pl^3}{EI}(\downarrow)$

Step 9: Determine the forces in the elements

- Element forces due to combined loads $\{p_C\}$

$$[p_C] = [T_{FA}]\{P_{CA}\}$$

$$= \begin{bmatrix} -2l & -1 \\ 0 & 1 \end{bmatrix} \begin{Bmatrix} -P/2 \\ Pl/4 \end{Bmatrix} = \begin{Bmatrix} 3Pl/4 \\ Pl/4 \end{Bmatrix}$$

- Element forces $\{p\}$

$$\{p\} = \{P_F\} + \{p_C\}$$

$$= \begin{Bmatrix} Pl/4 \\ -Pl/4 \end{Bmatrix} + \begin{Bmatrix} 3Pl/4 \\ Pl/4 \end{Bmatrix} = \begin{Bmatrix} Pl \\ 0 \end{Bmatrix}$$

- SFD and BMD

  Once $\{p\}$ is calculated, using the FBD of element (Figure 3.17i) the forces at the nodes (end actions) of the element are determined using equation of equilibrium. After finding the end actions, the SFD and BMD are drawn for the element.

$$\sum M_1 = 0 \Rightarrow Pl - Pl + 2lV_2 = 0 \Rightarrow V_2 = 0$$

$$\sum F_Y = 0 \Rightarrow V_1 - P = 0 \Rightarrow V_1 = P$$

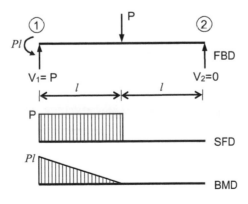

FIGURE 3.17i
SFD and BMD.

Example 3.10

Determine the vertical deflection at joint B of the beam (Figure 3.18a) and find the forces developed in the beam.

Solution

Step 1: Develop the discretized structure.
    The discretized structure is shown in Figure 3.18b.
Step 2: Determine the combined nodal loads
    The restrained structure, fixed end actions, and equivalent nodal loads are shown in Figures 3.18c and d. In this case, the combined nodal loads are same as equivalent nodal loads.
Step 3: Identify the global coordinates and define the load vector
    The global coordinates are shown in Figure 3.18e and the load vector is

$$\{P_{CA}\} = \left\{ \begin{array}{c} P_{C1} \\ P_{C2} \end{array} \right\} = \left\{ \begin{array}{c} -wl/2 \\ wl^2/12 \end{array} \right\}$$

Step 4: Identify the local coordinates and find $\{p_F\}$

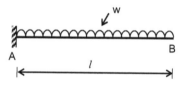

FIGURE 3.18a
Example 3.10.

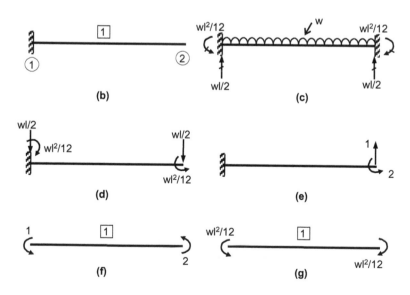

FIGURE 3.18b–g
Discretized structure, fixed end actions, and coordinates. (b) Discretized structure. (c) Fixed end actions and restrained structure. (d) Equivalent nodal loads. (e) Global Coordinates. (f) Combined local coordinates. (g) End moment in elements due to fixed end actions.

The combined local coordinates are shown in Figure 3.18f and the element forces due to fixed end actions (Figure 3.18g) are

$$\{P_F\} = \left\{ \begin{array}{c} wl^2 / 12 \\ -wl^2 / 12 \end{array} \right\}$$

Step 5: Develop the force transformation matrix $[T_{FA}]$

The end moments developed in the elements due to unit loads applied at global coordinates are shown in Figure 3.18h.

The force transformation matrix is

$$[T_{FA}] = \begin{array}{cc} {\scriptstyle P_{C1}=1} & {\scriptstyle P_{C2}=1} \\ \left[\begin{array}{c:c} -l & -1 \\ 0 & 1 \end{array} \right] \end{array}$$

Step 6: Develop $[f^i]$ and $[f]$

- Element flexibility matrix $[f^i]$

$$[f^1] = \frac{l}{6EI} \begin{bmatrix} 2 & -1 \\ -1 & 2 \end{bmatrix}$$

- Unassembled flexibility matrix

$$[f] = [f^1] = \frac{l}{6EI} \begin{bmatrix} 2 & -1 \\ -1 & 2 \end{bmatrix}$$

Step 7: Develop the flexibility matrix $[F_{AA}]$

$$[F_{AA}] = [T_{FA}]^T [f][T_{FA}]$$

$$= \frac{l}{6EI} \begin{bmatrix} 2l^2 & 3l \\ 3l & 6 \end{bmatrix}$$

Step 8: Determine the nodal displacements $\{D_A\}$

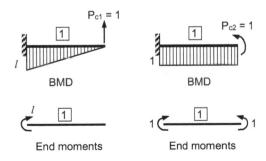

FIGURE 3.18h
Elements of $[T_{FA}]$.

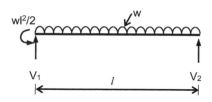

FIGURE 3.18i
FBD of element.

$$\{D_A\} = [F_{AA}]\{P_{CA}\}$$

$$\{D_A\} = \left\{ \begin{array}{c} D_1 \\ D_2 \end{array} \right\} = \left\{ \begin{array}{c} -wl^4/8EI \\ -wl^3/6EI \end{array} \right\}$$

Hence, the vertical deflection at joint B is $\dfrac{wl^4}{8EI}(\downarrow)$

Step 9: Determine the forces in the elements

- Element forces due to combined loads $\{p_C\}$

$$[p_C] = [T_{FA}]\{P_{CA}\}$$

$$= \left[\begin{array}{cc} -l & -1 \\ 0 & 1 \end{array} \right] \left\{ \begin{array}{c} -wl/2 \\ wl^2/12 \end{array} \right\} = \left\{ \begin{array}{c} 5wl^2/12 \\ wl^2/12 \end{array} \right\}$$

- Element forces $\{p\}$

$$\{p\} = \{P_F\} + \{p_C\}$$

$$= \left\{ \begin{array}{c} wl^2/12 \\ -wl^2/12 \end{array} \right\} + \left\{ \begin{array}{c} 5wl^2/12 \\ wl^2/12 \end{array} \right\} = \left\{ \begin{array}{c} wl^2/2 \\ 0 \end{array} \right\}$$

The FBD of the element is shown in Figure 3.18i.
  Using equations of equilibrium, it can be shown that
$V_1 = wl$ and $V_2 = 0$

Example 3.11

Find the deflection at joint C of the beam shown in Figure 3.19a. Also find the element forces.

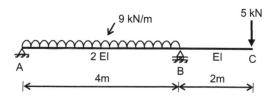

FIGURE 3.19a
Example 3.11.

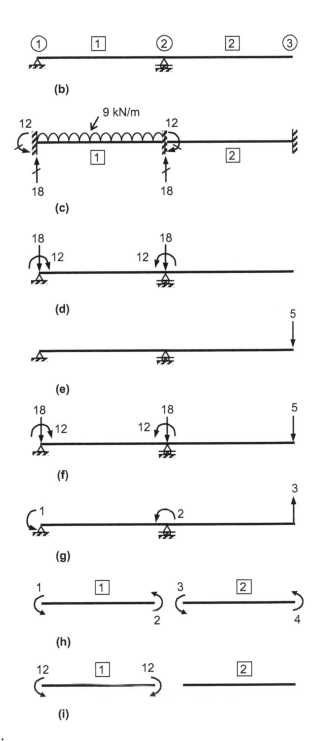

FIGURE 3.19b–i
Discretized structure, fixed end actions, and coordinates. (b) Discretized structure. (c) Restrained structure and fixed end actions. (d) Equivalent nodal loads. (e) Nodal loads. (f) Combined local coordinates (d) + (e). (g) Global coordinates. (h) Combined local coordinates. (i) End moment in elements due to fixed end actions.

Solution

Step 1: Develop the discretized structure.

The discretized structure is shown in Figure 3.19b.

Step 2: Determine the combined nodal loads

The restrained structure and the fixed end actions are shown in Figure 3.19c. The equivalent nodal loads are obtained by reversing the fixed end actions (Figure 3.19d). The actual nodal loads are shown in Figure 3.19e. The combined nodal loads are obtained by adding equivalent nodal loads with the actual nodal loads (Figure 3.19f).

Step 3: Identify the global coordinates and define the load vector

The global coordinates are shown in Figure 3.19g and the load vector is

$$\{P_{AC}\} = \begin{Bmatrix} P_{C1} \\ P_{C2} \\ P_{C3} \end{Bmatrix} = \begin{Bmatrix} -12 \\ 12 \\ -5 \end{Bmatrix}$$

Step 4: Identify the local coordinates and find $\{p_F\}$

The combined local coordinates are shown in Figure 3.19h and the end moments due to fixed end actions are indicated in Figure 3.19i. Hence,

$$\{p_F\} = \begin{Bmatrix} 12 \\ -12 \\ 0 \\ 0 \end{Bmatrix}$$

Step 5: Develop the force transformation matrix $[T_{FA}]$

The end moments developed in the elements due to unit loads applied at global coordinates are shown in Figure 3.19j.

From Figure 3.19j,

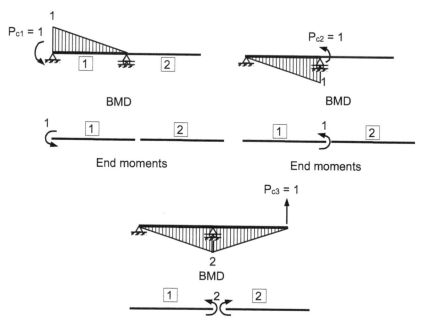

FIGURE 3.19j
Elements of $[T_{FA}]$.

$$[T_{FA}] = \begin{array}{ccc} P_{C1}=1 & P_{C2}=1 & P_{C3}=1 \\ \left[\begin{array}{c|c|c} 1 & 0 & 0 \\ 0 & 1 & 2 \\ 0 & 0 & -2 \\ 0 & 0 & 0 \end{array}\right] \end{array}$$

Step 6: Develop $[f^i]$ and $[f]$

- Element stiffness matrix $[f^i]$

$$[f^1] = \frac{4}{6(2EI)}\begin{bmatrix} 2 & -1 \\ -1 & 2 \end{bmatrix} = \frac{1}{6EI}\begin{bmatrix} 4 & -2 \\ -2 & 4 \end{bmatrix}$$

$$[f^2] = \frac{2}{6(EI)}\begin{bmatrix} 2 & -1 \\ -1 & 2 \end{bmatrix} = \frac{1}{6EI}\begin{bmatrix} 4 & -2 \\ -2 & 4 \end{bmatrix}$$

- Unassembled flexibility matrix

$$[f] = \begin{bmatrix} [f^1] & 0 \\ 0 & [f^2] \end{bmatrix} = \frac{1}{6EI}\begin{bmatrix} 4 & -2 & 0 & 0 \\ -2 & 4 & 0 & 0 \\ 0 & 0 & 4 & -2 \\ 0 & 0 & -2 & 4 \end{bmatrix}$$

Step 7: Develop the flexibility matrix $[F_{AA}]$

$$[F_{AA}] = [T_{FA}]^T[f][T_{FA}]$$

$$= \frac{1}{6EI}\begin{bmatrix} 4 & -2 & -4 \\ -2 & 4 & 8 \\ -4 & 8 & 32 \end{bmatrix}$$

Step 8: Determine the nodal displacements $\{D_A\}$

$$\{D_A\} = [F_{AA}]\{P_{CA}\}$$

$$\{D_A\} = \begin{Bmatrix} D_1 \\ D_2 \\ D_3 \end{Bmatrix} = \frac{1}{EI}\begin{Bmatrix} -8.67 \\ 5.33 \\ -2.67 \end{Bmatrix}$$

Step 9: Determine the forces in the elements

- Element forces due to combined loads $\{p_C\}$

$$[p_C] = [T_{FA}]\{P_{CA}\}$$

$$= \begin{bmatrix} -12 \\ 2 \\ 10 \\ 0 \end{bmatrix}$$

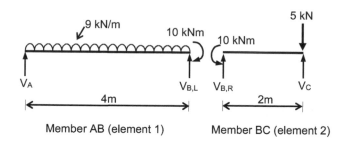

FIGURE 3.19k
FBD of members.

- Element forces $\{p\}$

$$\{p\} = \{P_F\} + \{p_C\}$$

$$= \left\{ \begin{array}{c} 0 \\ -10 \\ 10 \\ 0 \end{array} \right\}$$

The FBD of the two members are shown below (Figure 3.19k).
Using equations of equilibrium,
$V_A = 15.5$ kN, $V_{B,L} = 20.5$ kN, $V_{B,R} = 5$ kN, and $V_C = 0$ kN

3.9 Analysis of Statically Indeterminate Structures

In the case of statically indeterminate structures, the element forces $\{p\}$ and the action displacement vector $\{D_A\}$ can be calculated only after finding the redundant vector $\{P_X\}$. Hence, in the beginning stages of analysis, $\{P_X\}$ is determined using Eq. 3.19b.

$$\{D_X\} = [F_{XA}]\{P_A\} + [F_{XX}]\{P_X\}$$

$$\Rightarrow \quad \{P_X\} = [F_{XX}]^{-1}(\{D_X\} - [F_{XA}]\{P_A\}) \tag{3.27a}$$

If $\{D_X\} = \{0\}$, then

$$\{P_X\} = -[F_{XX}]^{-1}[F_{XA}]\{P_A\} \tag{3.27b}$$

The element forces $\{p\}$ are calculated using Eq. 3.8a.

$$\{p\} = [T_{FA}]\{P_A\} + [T_{FX}]\{P_X\}$$

The action displacement vector $\{D_A\}$ is determined using Eq. 3.19a.

$$\{D_A\}=[F_{AA}]\{P_A\}+[F_{AX}]\{P_X\}$$

The step-by-step procedure for the analysis of statically indeterminate structures is as follows:

i. Identify the nodes and elements, and draw the discretized structure.

ii. Choose the redundants and obtain the primary structure.

iii. Identify the global coordinates (action and redundant coordinates). Define the load vector $\{P_A\}$ and redundant vector $\{P_X\}$.

iv. Identify the local coordinates and combined local coordinates. Define the combined element force vector $\{p\}$.

v. Determine the force transformation matrices $[T_{FA}]$ and $[T_{FX}]$.

vi. Develop the individual element flexibility matrices $[f^1]$, $[f^2]$, ... and obtain the unassembled flexibility matrix $[f]$.

vii. Calculate $[F_{XA}]$ using the equation $[F_{XA}] = [T_{FX}]^T [f] [T_{FA}]$

viii. Calculate $[F_{XX}]$ using the equation $[F_{XX}] = [T_{FX}]^T [f] [T_{FX}]$

ix. Determine the redundant vector $\{P_X\} = [F_{XX}]^{-1} (\{D_X\} - [F_{XA}] \{P_A\})$

x. Calculate the element forces $\{p\}$.

$$\{p\}=[T_{FA}]\{P_A\}+[T_{FX}]\{P_X\}$$

xi. Calculate the nodal displacements $\{D_A\}$ using the equation

$$\{D_A\}=[F_{AA}]\{P_A\}+[F_{AX}]\{P_X\}$$

where $[F_{AX}] = [T_{FA}]^T [f] [T_{FX}] = [F_{XA}]^T$

3.9.1 Analysis of Structures Subjected to Element Loads

In the case of indeterminate structures, the combined force vector $\{P_C\}$ is partitioned as

$$\{P_C\} = \left\{ \begin{array}{c} \{P_{CA}\} \\ \hline \{P_{CX}\} \end{array} \right\}$$

where $\{P_{CA}\}$ and $\{P_{CX}\}$ are the combined forces along the action and redundant coordinates, respectively.

For the analysis of statically indeterminate structures subjected to element loads in the above steps, $\{P_{CA}\}$ and $\{P_{CX}\}$ are to be used instead of $\{P_A\}$ and $\{P_X\}$. The element forces caused by combined loads $\{p_C\}$ is determined as

$$\{p_C\}=[T_{FA}]\{P_{CA}\}+[T_{FX}]\{P_{CX}\} \tag{3.28}$$

After finding $\{p_C\}$, the element forces due to actual loads $\{p\}$ are calculated as

$$\{p\}=\{p_F\}+\{p_C\}$$

The nodal displacement $\{D_A\}$ are calculated using the following equation:

$$\{D_A\} = [F_{AA}]\{P_{CA}\} + [F_{AX}]\{P_{CX}\} \tag{3.29}$$

Example 3.12

Analyze the structure shown in Figure 3.20a.

Solution

Step 1: Develop the discretized structure.
   The discretized structure is shown in Figure 3.20b.
Step 2: Determine the combined nodal loads
   The restrained structure, fixed end actions, and equivalent nodal loads are shown in Figures 3.20c and d. In this case, the combined nodal loads are same as equivalent loads.
Step 3: Identify the global coordinates and define the load vector
   The reaction component B_y (Figure 3.20d) is taken as the redundant. The action coordinates, primary structure, redundant coordinates, and global coordinates are shown in Figures 3.20e–h. Figure 3.20i indicates the loads acting on the primary structure along the global coordinates. Hence,

- Action vector

$$\{P_{CA}\} = \{P_{C1}\} = \left\{\frac{wl^2}{12}\right\}$$

- Redundant vector

$$\{P_{CX}\} = \{P_{C2}\} = \left\{B_y - wl/2\right\}$$

Step 4: Identify the local coordinates and find $\{p_F\}$
   The combined local coordinates are shown in Figure 3.20j and the element forces due to fixed end actions are (Figure 3.20k)

$$\{p_F\} = \left\{\begin{array}{c} wl^2/12 \\ -wl^2/12 \end{array}\right\}$$

Step 5: Develop the force transformation matrix

- Elements of $[T_{FA}]$
     The end moments developed in the element due to $P_{C1} = 1$ are shown in Figure 3.20l.

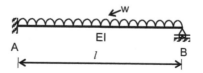

FIGURE 3.20a
Example 3.12.

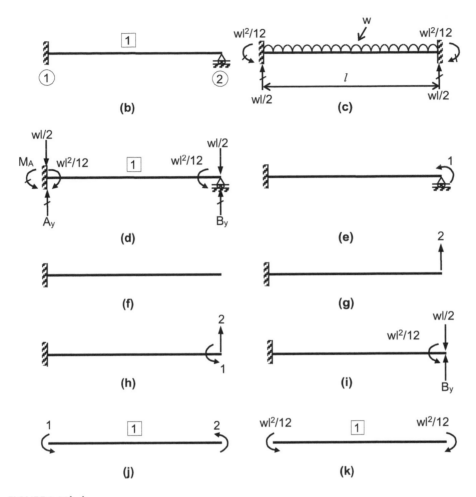

FIGURE 3.20b–k
Discretized structure, fixed end actions, and coordinates. (b) Discretized structure. (c) Restrained structure and fixed end actions. (d) Equivalent nodal loads. (e) Action coordinates. (f) Primary structure (DSI = 1). (g) Redundant coordinates. (h) Global coordinates. (i) Load acting on primary structure. (j) Combined local coordinates. (k) End moments due to fixed end actions.

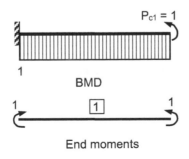

FIGURE 3.20l
Elements of $[T_{FA}]$.

Hence,

$$[T_{FA}] = \left\{ \begin{array}{c} -1 \\ 1 \end{array} \right\}$$

- Elements of $[T_{FX}]$
    From Figure 3.20m,

$$[T_{FX}] = \left\{ \begin{array}{c} -l \\ 0 \end{array} \right\}$$

Step 6: Develop $[f^i]$ and $[f]$

- Element stiffness matrix $[f^i]$

$$\left[f^1\right] = \frac{l}{6EI} \left[\begin{array}{cc} 2 & -1 \\ -1 & 2 \end{array} \right]$$

- Unassembled flexibility matrix

$$[f] = \left[f^1\right] = \frac{l}{6EI} \left[\begin{array}{cc} 2 & -1 \\ -1 & 2 \end{array} \right]$$

Step 7: Find $[F_{AA}], [F_{XX}], [F_{XA}], [F_{AX}]$

$$[F_{AA}] = [T_{FA}]^T [f][T_{FA}] = \frac{l}{EI}$$

$$[F_{XX}] = [T_{FX}]^T [f][T_{FX}] = \frac{l^3}{3EI}$$

$$[F_{XA}] = [T_{FX}]^T [f][T_{FA}] = \frac{l^2}{EI}$$

$$[F_{AX}] = [F_{XA}]^T = \frac{l^2}{2EI}$$

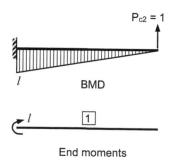

FIGURE 3.20m
Elements of $[T_{FX}]$.

Step 8: Determine the redundant vector $\{P_{CX}\}$

In this example, there is no settlement of support. Hence, $\{D_X\} = \{0\}$.

Hence,

$$\{P_{CX}\} = -[F_{XX}]^{-1}[F_{XA}]\{P_{CA}\}$$

$$\{P_{CX}\} = P_{C2} = -\frac{3EI}{l^3} \times \frac{l^2}{2EI} \times \frac{wl^2}{12} = -\frac{wl}{8}$$

Step 9: Determine the support reactions

From Step 8,

$$P_{C2} = B_y - \frac{wl}{2} = -\frac{wl}{8}$$

$$\Rightarrow \quad B_y = \frac{3}{8}wl$$

After finding B_y, the FBD shown in Figure 3.20d is used to find the other reaction components. From the figure,

$$\sum F_Y = 0 \Rightarrow A_y + B_y = wl \Rightarrow A_y = \frac{5}{8}wl$$

$$\sum M_1 = 0 \Rightarrow M_A - \frac{wl^2}{12} + \frac{wl^2}{12} + \left(\frac{3}{8}wl\right)l - \left(\frac{wl}{2}\right)l = 0$$

$$\Rightarrow M_A = \frac{wl^2}{8}$$

Step 10: Find the nodal displacements $\{D_A\}$

$$\{D_A\} = [F_{AA}]\{P_{CA}\} + [F_{AX}]\{P_{CX}\}$$

$$\{D_A\} = D_1 = \left(\frac{l}{EI} \times \frac{wl^2}{12}\right) + \left(\frac{l^2}{2EI} \times \left(\frac{-wl}{8}\right)\right) = \frac{wl^3}{48EI}$$

Step 11: Determine the forces in the elements

- Element forces due to combined loads $\{p_c\}$

$$\{p_c\} = [T_{FA}]\{P_{CA}\} + [T_{FX}]\{P_{CX}\}$$

$$= \left\{\begin{array}{c} -1 \\ 1 \end{array}\right\}\frac{wl^2}{12} + \left\{\begin{array}{c} -l \\ 0 \end{array}\right\}\left(\frac{-wl}{8}\right) = \left\{\begin{array}{c} wl^2/24 \\ wl^2/12 \end{array}\right\}$$

- Element forces $\{p\}$

$$\{p\} = \{p_F\} + \{p_C\}$$

$$= \left\{\begin{array}{c} wl^2/12 \\ -wl^2/12 \end{array}\right\} + \left\{\begin{array}{c} wl^2/24 \\ wl^2/12 \end{array}\right\} = \left\{\begin{array}{c} wl^2/8 \\ 0 \end{array}\right\}$$

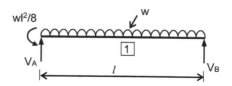

FIGURE 3.20n
FBD of element.

The FBD of the element is shown in Figure 3.20n.
   The end actions V_A and V_B are determined using equations of equilibrium. Thus,

$$V_A = \frac{5}{8}wl \text{ and } V_B = \frac{3}{8}wl$$

The SFD and BMD of the element can be drawn using the FBD.

Example 3.13

Find the forces developed in the structure shown in Figure 3.21a, if a rotation of $70/EI$ in the anticlockwise direction occurs at support A.

Solution

Step 1: Develop the discretized structure
   The discretized structure is shown in Figure 3.21b.
Step 2: Determine the combined nodal loads
   The combined nodal loads are shown in Figures 3.21d.
Step 3: Identify the global coordinates and define the force vector
   The support moment at A is taken as the redundant. The action coordinates, primary structure, redundant coordinates, and global coordinates are shown in Figures 3.21e–h. The corresponding loads are indicated in Figure 3.21i.

- Action vector

$$\{P_{CA}\} = \{P_{C1}\} = \{30\} \text{ kNm}$$

- Redundant vector

$$\{P_{CX}\} = \{P_{C2}\} = \{M_A - 30\} \text{ kNm}$$

Step 4: Identify the local coordinates and find $\{p_F\}$

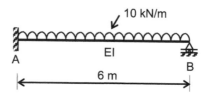

FIGURE 3.21a
Example 3.13.

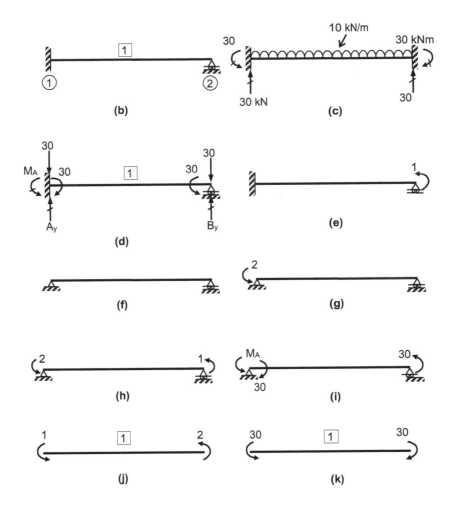

FIGURE 3.21b–k
Discretized structure, fixed end actions, and coordinates. (b) Discretized structure. (c) Restrained structure and fixed end actions. (d) Equivalent nodal loads and reactions. (e) Action coordinates. (f) Primary structure (DSI = 1). (g) Redundant coordinates. (h) Global coordinates. (i) Load acting on primary structure. (j) Combined local coordinates. (k) End moments due to fixed end actions.

The combined local coordinates are shown in Figure 3.21j. From Figure 3.21k,

$$\{p_F\} = \left\{ \begin{array}{c} 30 \\ -30 \end{array} \right\} \text{kNm}$$

Step 5: Develop $[T_{FA}]$ and $[T_{FX}]$

- $[T_{FA}]$
  From Figure 3.21l,

$$[T_{FA}] = \left\{ \begin{array}{c} 0 \\ 1 \end{array} \right\}$$

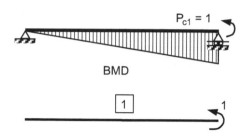

FIGURE 3.21l
Elements of $[T_{FA}]$.

- $[T_{FX}]$
  From Figure 3.21m,

$$[T_{FX}] = \left\{ \begin{array}{c} 1 \\ 0 \end{array} \right\}$$

Step 6: Develop $[f^i]$ and $[f]$

- Element stiffness matrix

$$[f^1] = \frac{6}{6EI} \begin{bmatrix} 2 & -1 \\ -1 & 2 \end{bmatrix} = \frac{1}{EI} \begin{bmatrix} 2 & -1 \\ -1 & 2 \end{bmatrix}$$

- Unassembled flexibility matrix

$$[f] = [f^1] = \frac{1}{EI} \begin{bmatrix} 2 & -1 \\ -1 & 2 \end{bmatrix}$$

Step 7: Find $[F_{XX}]$, $[F_{XA}]$

$$[F_{XX}] = [T_{FX}]^T [f][T_{FX}] = 2/EI$$

$$[F_{XA}] = [T_{FX}]^T [f][T_{FA}] = -1/EI$$

Step 8: Determine the redundant vector $\{P_{CX}\}$

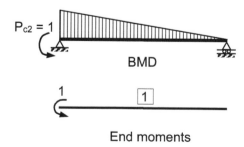

FIGURE 3.21m
Elements of $[T_{FX}]$.

$$\{D_X\} = D_2 = \frac{70}{EI}$$

$$\{P_{CX}\} = [F_{XX}]^{-1}(\{D_X\} - [F_{XA}]\{P_{CA}\})$$

$$\therefore \{P_{CX}\} = P_{C2} = \frac{EI}{2}\left(\frac{70}{EI} + \frac{30}{EI}\right) = 50 \text{ kNm}$$

Step 9: Determine the forces in the elements

- Element forces due to $\{P_C\}$

$$\{p_C\} = [T_{FA}]\{P_{CA}\} + [T_{FX}]\{P_{CX}\} = \begin{Bmatrix} 50 \\ 30 \end{Bmatrix} \text{ kNm}$$

- Element forces $\{p\}$

$$\{p\} = \{p_F\} + \{p_C\} = \begin{Bmatrix} 80 \\ 0 \end{Bmatrix} \text{ kNm}$$

Once the end moments are obtained, the other end actions are determined using the FBD of the member.

Alternate Method

Alternatively the support settlements can be converted into fixed end actions (Figure 3.21n).

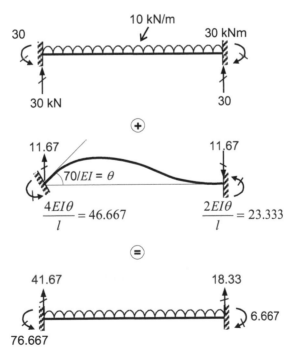

FIGURE 3.21n
Fixed end actions.

FIGURE 3.21o
Combined nodal loads.

Thus,

$$\{p_F\} = \left\{ \begin{array}{c} 76.667 \\ -6.667 \end{array} \right\}$$

The combined nodal loads are shown in Figure 3.21o.

$$\{P_{CA}\} = P_{C1} = 6.667$$

$$\{P_{CX}\} = P_{C2} = M_A - 76.667$$

Since the effect of support settlement is included in combined nodal loads, $\{D_X\} = \{0\}$. Hence, the redundant vector is

$$\{P_{CX}\} = -[F_{XX}]^{-1}[F_{XA}]\{P_{CA}\}$$

$$\{P_{CX}\} = P_{C2} = 3.334 \text{ kNm}$$

The element force vector is

$$\{p\} = \{p_F\} + [T_{FA}]\{P_{CA}\} + [T_{FX}]\{P_{CX}\} = \left\{ \begin{array}{c} 80 \\ 0 \end{array} \right\} \text{kNm}$$

Example 3.14

Find the forces developed in the beam shown in Figure 3.21a, if support A rotates by $70/EI$ in the anticlockwise direction and support B sinks by $100/EI$.

Solution

Step 1: Develop the discretized structure
   The discretized structure is shown in Figure 3.21b.
Step 2: Determine the combined nodal loads.
   The support settlements are converted into fixed end actions (Figure 3.22a) and the combined nodal loads are shown in Figure 3.22b.
Step 3: Identify the global coordinates and define the force vector
   The global coordinates are shown in Figure 3.21h. The loads acting on the primary structure are indicated in Figure 3.22c.

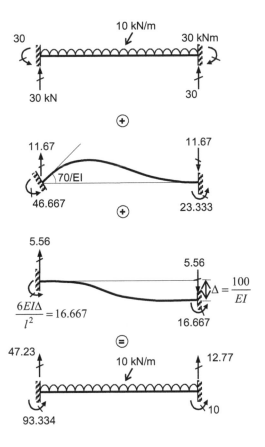

FIGURE 3.22a
Fixed end actions.

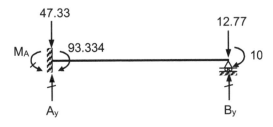

FIGURE 3.22b
Combined nodal loads and reactions.

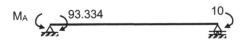

FIGURE 3.22c
Loads on primary structure.

- Action vector

$$\{P_{CA}\} = P_{C1} = -10 \text{ kNm}$$

- Redundant vector

$$\{P_{CX}\} = P_{C2} = M_A - 93.334 \text{ kNm}$$

Step 4: Identify the local coordinates and find $\{p_F\}$
The combined local coordinates are shown in Figure 3.21j.
From Figure 3.22d,

$$\{p_F\} = \left\{ \begin{array}{c} 93.334 \\ 10 \end{array} \right\} \text{kNm}$$

Step 5: Develop $[T_{FA}]$ and $[T_{FX}]$
From Figures 3.21l and m,

$$[T_{FA}] = \left\{ \begin{array}{c} 0 \\ 1 \end{array} \right\}$$

$$[T_{FX}] = \left\{ \begin{array}{c} 1 \\ 0 \end{array} \right\}$$

Step 6: Develop $[f^i]$ and $[f]$

- Element stiffness matrix

$$[f^1] = \frac{1}{EI} \left[\begin{array}{cc} 2 & -1 \\ -1 & 2 \end{array} \right]$$

$$[f] = \frac{1}{EI} \left[\begin{array}{cc} 2 & -1 \\ -1 & 2 \end{array} \right]$$

Step 7: Find $[F_{XX}]$, $[F_{XA}]$

$$[F_{XX}] = 2/EI$$

$$[F_{XA}] = -1/EI$$

Step 8: Determine the redundant vector $\{P_{CX}\}$

$$\{D_X\} = \{0\}$$

FIGURE 3.22d
End moments due to fixed end actions.

$$\{P_{CX}\} = -[F_{XX}]^{-1}[F_{XA}]\{P_{CA}\}$$

$$\{P_{CX}\} = P_{C2} = -5 \text{ kNm}$$

Step 9: Determine the element forces vector $\{p\}$

$$\{p\} = \{p_F\} + [T_{FA}]\{P_{CA}\} + [T_{FX}]\{P_{CX}\} = \left\{ \begin{array}{c} 88.334 \\ 0 \end{array} \right\} \text{kNm}$$

Example 3.15

Find the forces developed in the beam shown in Figure 3.23a, if support B and C sinks by 100/*EI* and 50/*EI*, respectively.

Solution

Step 1: Develop the discretized structure

The nodes and elements are shown in Figure 3.23b. The loads are treated as element loads.

Step 2: Determine the combined nodal loads

The restrained structure and the fixed end actions for individual elements and structure are shown in Figures 3.23c–e. The combined loads are shown in Figure 3.23f.

Step 3: Identify the global coordinates and force vector

The action coordinates, primary structure, redundant coordinates, and the global coordinates are shown in Figure 3.23g–j. The vertical reactions at supports B and C are chosen as the redundant.

From Figure 3.23k,

$$\{P_{CA}\} = \left\{ \begin{array}{c} P_{C1} \\ P_{C2} \end{array} \right\} = \left\{ \begin{array}{c} 5 \\ 5 \end{array} \right\}$$

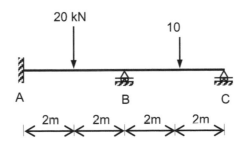

FIGURE 3.23a
Example 3.15.

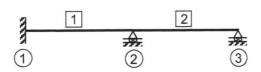

FIGURE 3.23b
Discretized structure.

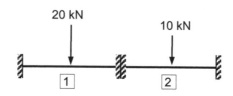

FIGURE 3.23c
Restrained structure.

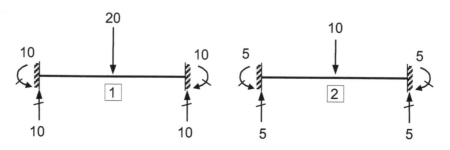

FIGURE 3.23d
Fixed end actions of individual elements.

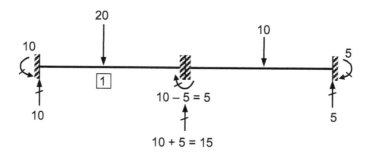

FIGURE 3.23e
Fixed end actions of the structure.

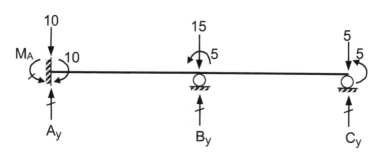

FIGURE 3.23f
Equivalent nodal loads and reactions.

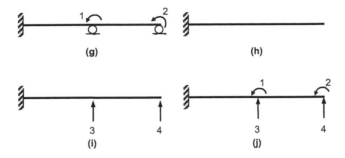

FIGURE 3.23g–j
Primary structure and coordinates. (g) Action coordinates. (h) Primary structure. (i) Redundant coordinates. (j) Global coordinates.

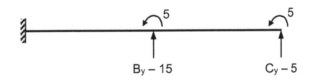

FIGURE 3.23k
Loads on primary structure.

$$\{P_{CX}\} = \left\{ \begin{array}{c} P_{C3} \\ P_{C4} \end{array} \right\} = \left\{ \begin{array}{c} B_y - 15 \\ C_y - 5 \end{array} \right\}$$

Step 4: Identify the local coordinates and find $\{p_F\}$

The combined local coordinates are indicated in Figure 3.23l. The element forces due to fixed end actions (Figure 3.23d) are shown in Figure 3.23m. Thus,

FIGURE 3.23l
Combined local coordinates.

FIGURE 3.23m
End moments due to fixed end actions.

$$\{p_F\} = \left\{ \begin{array}{c} 10 \\ -10 \\ 5 \\ -5 \end{array} \right\}$$

Step 5: Develop $[T_{FA}]$ and $[T_{FX}]$
From Figure 3.23n,

$$[T_{FA}] = \left[\begin{array}{cc} -1 & -1 \\ 1 & 1 \\ 0 & -1 \\ 0 & 1 \end{array} \right]$$

From Figure 3.23o,

$$[T_{FX}] = \left[\begin{array}{cc} -4 & -8 \\ 0 & 4 \\ 0 & -4 \\ 0 & 0 \end{array} \right]$$

Step 6: Develop $[f^i]$ and $[f]$

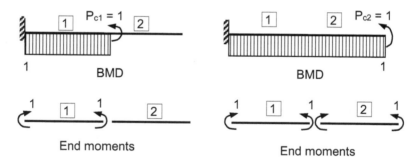

FIGURE 3.23n
Elements of $[T_{FA}]$.

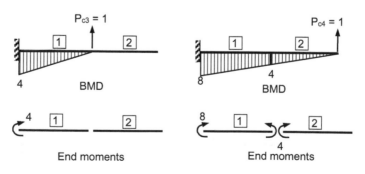

FIGURE 3.23o
Elements of $[T_{FX}]$.

- Element stiffness matrix

$$[f^1] = [f^2] = \frac{4}{6EI} \begin{bmatrix} 2 & -1 \\ -1 & 2 \end{bmatrix} = \frac{2}{3EI} \begin{bmatrix} 2 & -1 \\ -1 & 2 \end{bmatrix}$$

- Unassembled flexibility matrix

$$[f] = \frac{2}{3EI} \begin{bmatrix} 2 & -1 & 0 & 0 \\ -1 & 2 & 0 & 0 \\ 0 & 0 & 2 & -1 \\ 0 & 0 & -1 & 2 \end{bmatrix}$$

Step 7: Find $[F_{XX}]$, $[F_{XA}]$

$$[F_{XX}] = \frac{1}{EI} \begin{bmatrix} 21.33 & 53.33 \\ 53.33 & 170.67 \end{bmatrix}$$

$$[F_{XA}] = \frac{1}{EI} \begin{bmatrix} 8 & 8 \\ 24 & 32 \end{bmatrix}$$

Step 8: Determine $\{P_{CX}\}$

$$D_X = \begin{Bmatrix} D_3 \\ D_4 \end{Bmatrix} = \frac{1}{EI} \begin{Bmatrix} -100 \\ -50 \end{Bmatrix}$$

$$\{P_{CX}\} = [F_{XX}]^{-1}(\{D_X\} - [F_{XA}]\{P_{CA}\})$$

$$\{P_{CX}\} = \begin{Bmatrix} P_{C3} \\ P_{C4} \end{Bmatrix} = \begin{Bmatrix} -16.473 \\ 3.214 \end{Bmatrix}$$

Step 9: Determine the end moments in elements $\{p\}$

$$\{p\} = \{p_F\} + [T_{FA}]\{P_{CA}\} + [T_{FX}]\{P_{CX}\} = \begin{Bmatrix} 40.179 \\ 12.857 \\ -12.857 \\ 0 \end{Bmatrix} \text{kNm}$$

Alternate Method

The support settlement can be converted into fixed end actions. Figure 3.23p shows the relative settlements (Δ_{AB} and Δ_{BC}) of the two members of the beam.

The fixed end actions due to relative settlements for the two elements are indicated in Figure 3.23q.

To this, the fixed end actions due to the element loads (Figure 3.23d) are added and the total fixed end actions of the elements and structure are shown in Figures 3.23r and s.

The combined nodal loads and the support reactions are shown in Figure 3.23t.

The loads corresponding to the global coordinates indicated in Figure 3.23j are

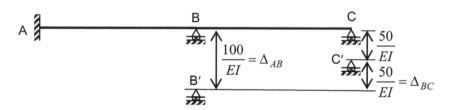

FIGURE 3.23p
Relative settlements of the members.

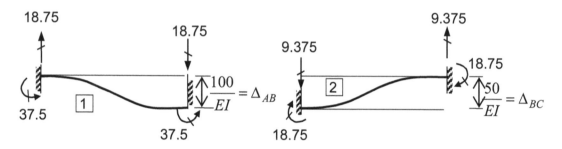

FIGURE 3.23q
Fixed end actions due to relative settlements.

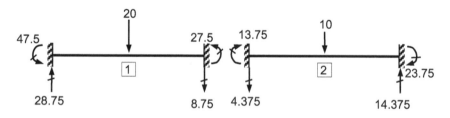

FIGURE 3.23r
Fixed end actions of elements.

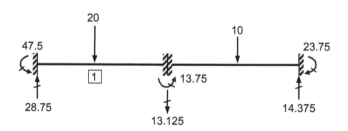

FIGURE 3.23s
Fixed end actions of structure.

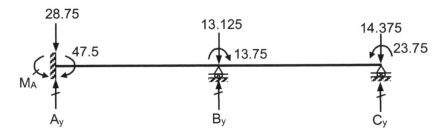

FIGURE 3.23t
Combined nodal loads and reactions.

$$\{P_{CA}\} = \left\{ \begin{array}{c} P_{C1} \\ P_{C2} \end{array} \right\} = \left\{ \begin{array}{c} -13.75 \\ 23.75 \end{array} \right\}$$

$$\{P_{CX}\} = \left\{ \begin{array}{c} P_{C3} \\ P_{C4} \end{array} \right\} = \left\{ \begin{array}{c} B_y - 13.125 \\ C_y - 14.375 \end{array} \right\}$$

The element force vector due to fixed end actions (Figure 3.23r) is

$$\{P_F\} = \left\{ \begin{array}{c} 47.5 \\ 27.5 \\ -13.75 \\ -23.75 \end{array} \right\} \text{kNm}$$

Since the support settlements are converted into fixed end actions, $\{D_X\} = \{0\}$. Thus, the redundant force vector is

$$\{P_{CX}\} = -[F_{XX}]^{-1}[F_{XA}]\{P_{CA}\}$$

$$= \left\{ \begin{array}{c} 11.652 \\ -6.161 \end{array} \right\} \text{kN}$$

and the end moments in the elements are

$$\{p\} = \{p_F\} + [T_{FA}]\{P_{CA}\} + [T_{FX}]\{P_{CX}\}$$

$$= \left\{ \begin{array}{c} 40.179 \\ 12.857 \\ -12.857 \\ 0 \end{array} \right\} \text{kNm}$$

Example 3.16

Determine the forces developed in the beam shown in Figure 3.24a.

Solution

Step 1: Develop the discretized structure.

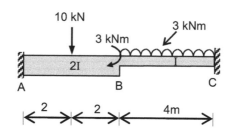

FIGURE 3.24a
Example 3.16.

The beam is modeled using two elements as shown in Figure 3.24b.
Step 2: Determine the combined nodal loads.
   The applied loads are grouped into nodal loads and element loads (Figure 3.24c). The element loads are applied on the restrained structure (Figure 3.24d) and the fixed end actions are determined (Figures 3.24e and f). The equivalent nodal loads are shown in Figure 3.24g. To this, the nodal load (Figure 3.24c) is added to get the combined nodal loads (Figure 3.24h).
Step 3: Identify the global coordinates and force vector
   The global coordinates are shown in Figure 3.24i. The reactions at support C (C_y and M_c) are taken as the redundants.
   From Figure 3.24j,

$$\{P_{CA}\} = \left\{ \begin{array}{c} P_{C1} \\ P_{C2} \end{array} \right\} = \left\{ \begin{array}{c} -11 \\ -2 \end{array} \right\}$$

$$\{P_{CX}\} = \left\{ \begin{array}{c} P_{C3} \\ P_{C4} \end{array} \right\} = \left\{ \begin{array}{c} C_y - 6 \\ 4 - M_c \end{array} \right\}$$

Step 4: Identify the local coordinates and find $\{p_F\}$.
   The combined local coordinates are shown in Figure 3.24k.
   From Figures 3.24 e and l,

$$\{p_F\} = \left\{ \begin{array}{c} 5 \\ -5 \\ 4 \\ -4 \end{array} \right\} \text{kNm}$$

Step 5: Develop $[T_{FA}]$ and $[T_{FX}]$

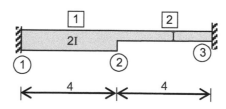

FIGURE 3.24b
Discretized structure.

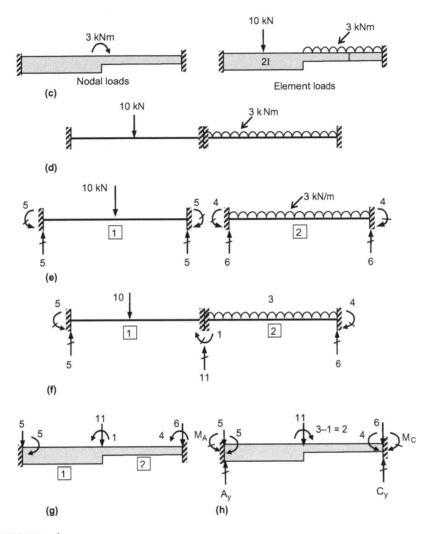

FIGURE 3.24c–h
Loads and fixed end actions. (c) Load on beam. (d) Restrained structure subjected to element loads.
(e) Fixed end actions of elements. (f) Fixed end actions for structure. (g) Equivalent nodal loads.
(h) combined nodal loads and reactions.

From Figures 3.24m and n,

$$[T_{FA}] = \begin{bmatrix} -4 & -1 \\ 0 & 1 \\ 0 & 0 \\ 0 & 0 \end{bmatrix} \quad [T_{FX}] = \begin{bmatrix} -8 & -1 \\ 4 & 1 \\ -4 & -1 \\ 0 & 1 \end{bmatrix}$$

Step 6: Develop $[f^i]$ and $[f]$

- Element stiffness matrix

$$[f^1] = \frac{4}{6(2EI)} \begin{bmatrix} 2 & -1 \\ -1 & 2 \end{bmatrix} = \frac{1}{6EI} \begin{bmatrix} 4 & -2 \\ -2 & 4 \end{bmatrix}$$

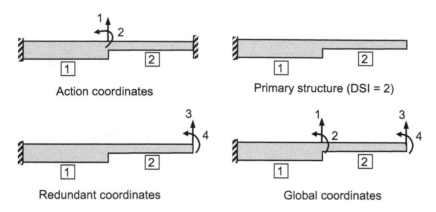

FIGURE 3.24i
Primary structure and global coordinates.

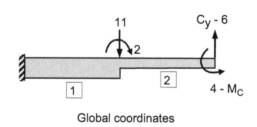

FIGURE 3.24j
Loads on primary structure.

FIGURE 3.24k
Combined local coordinates.

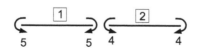

FIGURE 3.24l
Elements of $\{p_F\}$.

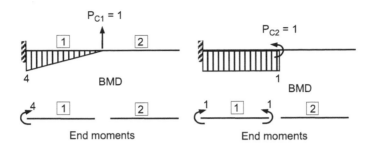

FIGURE 3.24m
Elements of $[T_{FA}]$.

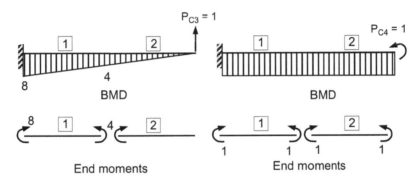

FIGURE 3.24n
Elements of $[T_{FX}]$.

$$\left[f^2\right] = \frac{4}{6EI}\begin{bmatrix} 2 & -1 \\ -1 & 2 \end{bmatrix} = \frac{1}{6EI}\begin{bmatrix} 8 & -4 \\ -4 & 8 \end{bmatrix}$$

- Unassembled flexibility matrix

$$[f] = \begin{bmatrix} [f^1] & 0 \\ 0 & [f^2] \end{bmatrix} = \frac{1}{6EI}\begin{bmatrix} 4 & -2 & 0 & 0 \\ -2 & 4 & 0 & 0 \\ 0 & 0 & 8 & -4 \\ 0 & 0 & -4 & 8 \end{bmatrix}$$

Step 7: Find $[F_{XX}]$, $[F_{XA}]$

$$[F_{XX}] = \frac{1}{EI}\begin{bmatrix} 96 & 20 \\ 20 & 6 \end{bmatrix}$$

$$[F_{XA}] = \frac{1}{EI}\begin{bmatrix} 26.667 & 12 \\ 4 & 2 \end{bmatrix}$$

Step 8: Determine $\{P_{CX}\}$

$$\{D_X\} = \{0\}$$

$$\{P_{CX}\} = \left\{ \begin{array}{c} P_{C3} \\ P_{C4} \end{array} \right\} = \left\{ \begin{array}{c} 5.364 \\ -9.879 \end{array} \right\}$$

Step 9: Determine the end moments in elements $\{p\}$

$$\{p\} = \{p_F\} + [T_{FA}]\{P_{CA}\} + [T_{FX}]\{P_{CX}\} = \left\{ \begin{array}{c} 17.97 \\ 4.576 \\ -7.576 \\ -13.879 \end{array} \right\} \text{kNm}$$

Example 3.17

Determine the forces developed in the plane frame shown in Figure 3.25a.

Solution

Step 1: Develop the discretized structure.
   The elements and nodes are shown in Figure 3.25b.
Step 2: Determine the combined nodal loads.
   The loads, restrained structure, fixed end actions, equivalent nodal loads, and combined nodal loads are shown in Figures 3.25c–h.
Step 3: Identify the global coordinates and force vector.

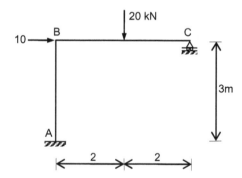

FIGURE 3.25a
Example 3.17.

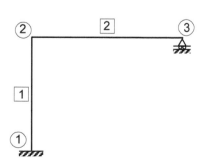

FIGURE 3.25b
Discretized structure.

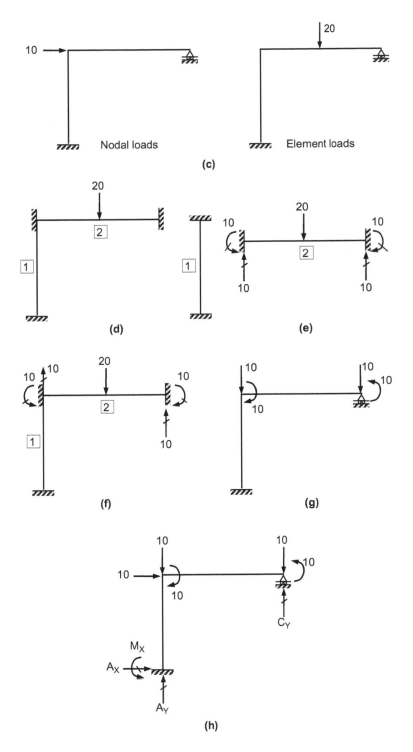

FIGURE 3.25c–h
Loads and fixed end actions. (c) Loads on the structure. (d) Restrained structure. (e) Fixed end actions of elements. (f) Fixed end actions for structure. (g) Equivalents nodal loads. (h) Combined nodal loads and reactions.

The global coordinates and the corresponding loads are shown in Figures 3.25i and j. The force vector is

$$\{P_{CA}\} = \begin{Bmatrix} P_{C1} \\ P_{C2} \\ P_{C3} \end{Bmatrix} = \begin{Bmatrix} 10 \\ -10 \\ 10 \end{Bmatrix}$$

$$\{P_{CX}\} = \{P_{C4}\} = \{C_y - 10\}$$

Step 4: Identify the local coordinates and find $\{p_F\}$.
  The combined local coordinates are indicated in Figure 3.25k.
  From Figures 3.25e and l,

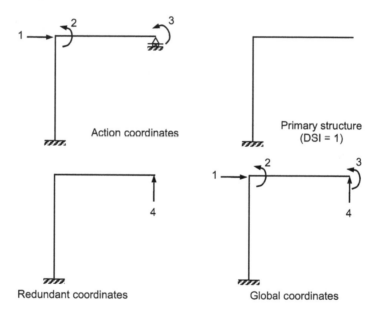

Action coordinates

Primary structure
(DSI = 1)

Redundant coordinates

Global coordinates

FIGURE 3.25i
Primary structure and global coordinates.

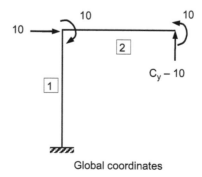

Global coordinates

FIGURE 3.25j
Loads on primary structure.

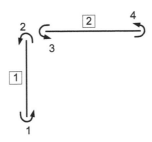

FIGURE 3.25k
Combined local coordinates.

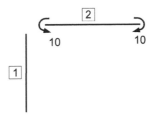

FIGURE 3.25l
Elements of $\{p_F\}$.

$$\{p_F\} = \begin{Bmatrix} 0 \\ 0 \\ 10 \\ -10 \end{Bmatrix}$$

Step 5: Develop $[T_{FA}]$ and $[T_{FX}]$
From Figures 3.25m and n,

$$[T_{FA}] = \begin{bmatrix} 3 & -1 & -1 \\ 0 & 1 & 1 \\ 0 & 0 & -1 \\ 0 & 0 & 1 \end{bmatrix} \text{ and } [T_{FX}] = \begin{bmatrix} -4 \\ 4 \\ -4 \\ 0 \end{bmatrix}$$

Step 6: Develop $[f^i]$ and $[f]$

- Element stiffness matrix

$$\left[f^1\right] = \frac{3}{6EI}\begin{bmatrix} 2 & -1 \\ -1 & 2 \end{bmatrix} = \frac{1}{6EI}\begin{bmatrix} 6 & -3 \\ -3 & 6 \end{bmatrix}$$

$$\left[f^2\right] = \frac{4}{6EI}\begin{bmatrix} 2 & -1 \\ -1 & 2 \end{bmatrix} = \frac{1}{6EI}\begin{bmatrix} 8 & -4 \\ -4 & 8 \end{bmatrix}$$

- Unassembled flexibility matrix

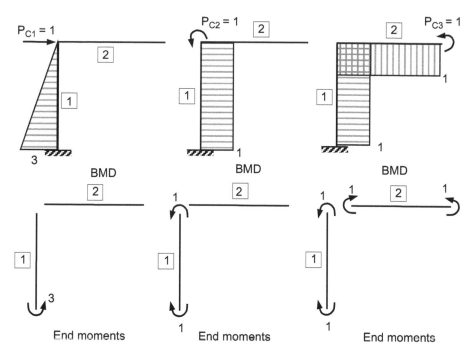

FIGURE 3.25m
Elements of $[T_{FA}]$.

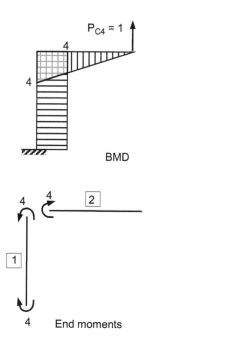

FIGURE 3.25n
Elements of $[T_{FX}]$.

$$[f] = \frac{1}{6EI} \begin{bmatrix} 6 & -3 & 0 & 0 \\ -3 & 6 & 0 & 0 \\ 0 & 0 & 8 & -4 \\ 0 & 0 & -4 & 8 \end{bmatrix}$$

Step 7: Find $[F_{XX}]$, $[F_{XA}]$

$$[F_{XX}] = \left[\frac{69.333}{EI} \right]$$

$$[F_{XA}] = \frac{1}{EI} \begin{bmatrix} -18 & 12 & 30 \end{bmatrix}$$

Step 8: Determine $\{P_{CX}\}$

$$\{D_X\} = \{0\}$$

$$\{P_{CX}\} = P_{C4} = 1.44 \text{ kN}$$

Step 9: Determine the end moments in elements $\{p\}$

$$\{p\} = \{p_F\} + [T_{FA}]\{P_{CA}\} + [T_{FX}]\{P_{CX}\} = \begin{Bmatrix} 24.231 \\ 5.769 \\ -5.769 \\ 0 \end{Bmatrix} \text{kNm}$$

Step 10: Determine the forces in the elements
The FBD of members and joints are shown in Figure 3.25o.

- FBD of member BC

$$\sum M_B = 0 \Rightarrow 4V_C = (2 \times 20) + 5.679 \Rightarrow V_C = 11.42 \text{ kN}$$

$$\sum F_Y = 0 \Rightarrow V_{BC} = 20 - V_C = 8.58 \text{ kN}$$

$$\sum F_X = 0 \Rightarrow N_{BC} + N_C = 0$$

- FBD of joint C

$$\sum F_X = 0 \Rightarrow N_C = 0 \Rightarrow N_{BC} = 0$$

$$\sum F_Y = 0 \Rightarrow C_Y = V_C = 11.42 \text{ kN}$$

- FBD of joint B

$$\sum F_X = 0 \Rightarrow V_{BA} + 10 - N_{BC} = 0$$

$$\because N_{BC} = 0, V_{BA} = -10 \text{ kN}$$

$$\sum F_Y = 0 \Rightarrow N_{BA} + V_{BC} = 0 \Rightarrow N_{BA} = -V_{BC} = -8.58 \text{ kN}$$

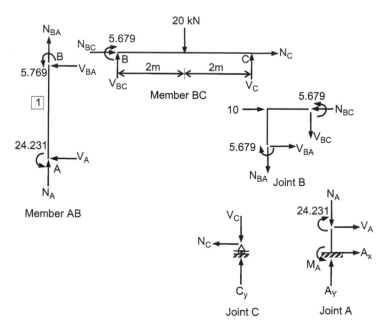

FIGURE 3.25o
FBD of joints and members.

- FBD of member AB

$$\sum F_X = 0 \Rightarrow V_A = -V_{BA} = 10 \text{ kN}$$

$$\sum F_Y = 0 \Rightarrow N_A + N_{BA} = 0 \Rightarrow N_A = -N_{BA} = 8.58 \text{ kN}$$

- FBD of joint A

$$\sum F_Y = 0 \Rightarrow A_Y = N_A = 8.58 \text{ kN}$$

$$\sum F_X = 0 \Rightarrow A_X = -V_A = -10 \text{ kN}$$

$$\sum M_A = 0 \Rightarrow M_A = 24.231 \text{ kNm}$$

Thus, the end actions in members and support reactions can be determined using the FBD of members and joints. Even if beam element is used to model the plane frame structure, the axial forces developed in the members are found using equations of equilibrium.

Example 3.18

Determine the forces developed in the plane frame shown in Figure 3.26a.

Solution

Step 1: Develop the discretized structure.
  The discretized structure is shown in Figure 3.26b.

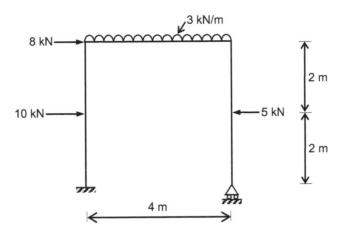

FIGURE 3.26a
Example 3.18.

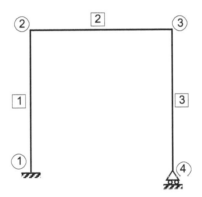

FIGURE 3.26b
Discretized structure.

Step 2: Determine the combined nodal loads.
   The loads, fixed end actions, equivalent nodal loads, and combined nodal loads are shown in Figures 3.26c–h.
Step 3: Identify the global coordinates and force vector
   From Figures 3.26i and j,

$$\{P_{CA}\} = \begin{Bmatrix} P_{C1} \\ P_{C2} \\ P_{C3} \\ P_{C4} \\ P_{C5} \end{Bmatrix} = \begin{Bmatrix} 1 \\ 1.5 \\ 10.5 \\ -2.5 \\ 2.5 \end{Bmatrix} \text{ and } \{P_{CX}\} = \{P_{C6}\} = \{D_y - 6\}$$

Step 4: Identify the local coordinates and find $\{p_F\}$.
   The combined local coordinates are indicated in Figure 3.26k.
   From Figures 3.26e and l,

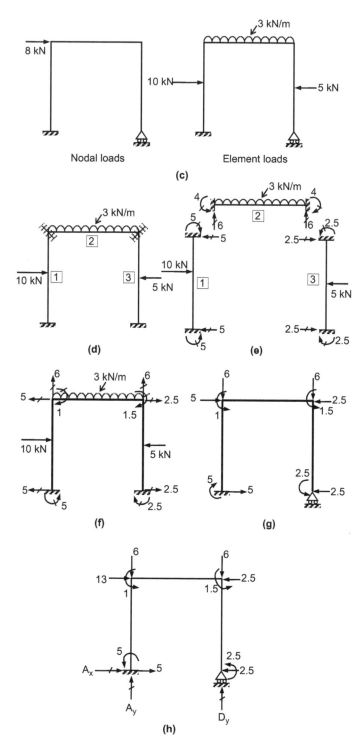

FIGURE 3.26c–h

Loads and fixed end actions. (c) Loads on structure. (d) Restrained structure. (e) Fixed end action of elements. (f) Fixed end action of structure. (g) Equivalent nodal loads. (h) Combined nodal loads.

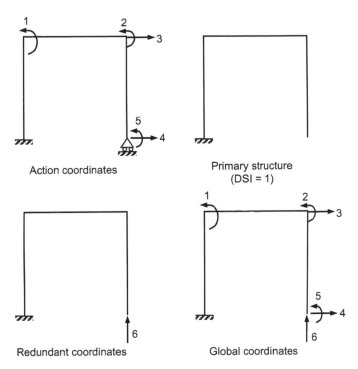

FIGURE 3.26i
Primary structure and global coordinates.

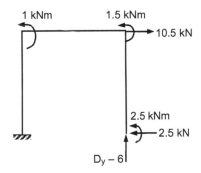

FIGURE 3.26j
Loads on primary structure.

$$\{p_F\} = \begin{Bmatrix} 5 \\ -5 \\ 4 \\ -4 \\ 2.5 \\ -2.5 \end{Bmatrix}$$

Step 5: Develop $[T_{FA}]$ and $[T_{FX}]$
From Figures 3.26m and n,

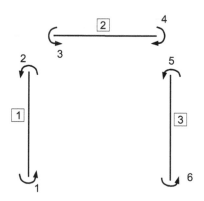

FIGURE 3.26k
Combined local coordinates.

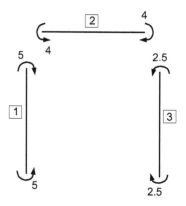

FIGURE 3.26l
Elements of $\{p_F\}$.

$$[T_{FA}] = \begin{bmatrix} -1 & -1 & 4 & 0 & -1 \\ 1 & 1 & 0 & 4 & 1 \\ 0 & -1 & 0 & -4 & -1 \\ 0 & 1 & 0 & 4 & 1 \\ 0 & 0 & 0 & -4 & -1 \\ 0 & 0 & 0 & 0 & 1 \end{bmatrix} \text{ and } [T_{FX}] = \begin{bmatrix} -4 \\ 4 \\ -4 \\ 0 \\ 0 \\ 0 \end{bmatrix}$$

Step 6: Develop $[f^i]$ and $[f]$

- Element stiffness matrix

$$\left[f^1\right] = \left[f^2\right] = \left[f^3\right] = \frac{4}{6EI}\begin{bmatrix} 2 & -1 \\ -1 & 2 \end{bmatrix} = \frac{2}{3EI}\begin{bmatrix} 6 & -3 \\ -3 & 6 \end{bmatrix}$$

- Unassembled flexibility matrix

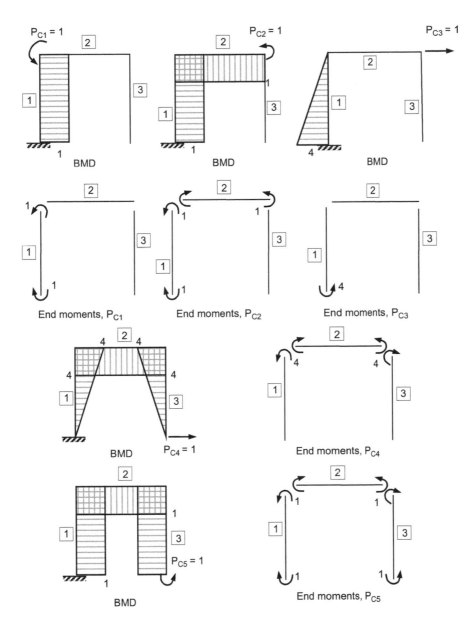

FIGURE 3.26m
Elements of $[T_{FA}]$.

$$[f] = \frac{2}{3EI} \begin{bmatrix} 2 & -1 & 0 & 0 & 0 & 0 \\ -1 & 2 & 0 & 0 & 0 & 0 \\ 0 & 0 & 2 & -1 & 0 & 0 \\ 0 & 0 & -1 & 2 & 0 & 0 \\ 0 & 0 & 0 & 0 & 2 & -1 \\ 0 & 0 & 0 & 0 & -1 & 2 \end{bmatrix}$$

Step 7: Find $[F_{XX}]$, $[F_{XA}]$

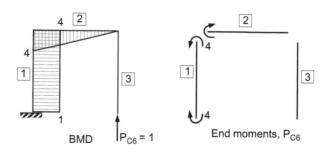

FIGURE 3.26n
Elements of $[T_{FX}]$.

$$[F_{XX}] = \left[\frac{85.333}{EI} \right]$$

$$[F_{XA}] = \frac{1}{EI} \begin{bmatrix} 16 & 24 & -32 & 64 & 24 \end{bmatrix}$$

Step 8: Determine $\{P_{CX}\}$

$$\{D_X\} = \{0\}$$

$$\{P_{CX}\} = P_{C6} = 4.5 \text{ kN}$$

Step 9: Determine the end moments in elements $\{p\}$

$$\{p\} = \begin{Bmatrix} 24 \\ 8 \\ -8 \\ -10 \\ 10 \\ 0 \end{Bmatrix} \text{kNm}$$

Example 3.19

Find the forces developed in the truss shown in Figure 3.27a. Also find the horizontal and vertical displacements at joint A. The stiffness of all the members is $(AE/l)50$ kN/mm.

Solution

Step 1: Develop the discretized structure.
   The discretized structure is shown in Figure 3.27b.
Step 2: Identify the global coordinates and force vector
   DSI of the structure is 1. The force in the member AC (P_{AC}) is taken as the redundant. The global coordinates, primary structure, and loads are shown in Figures 3.27c–g. From Figure 3.27g,

$$\{P_A\} = \begin{Bmatrix} P_1 \\ P_2 \end{Bmatrix} = \begin{Bmatrix} 0 \\ -10 \end{Bmatrix} \text{kN}$$

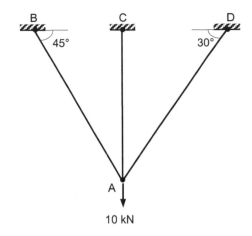

FIGURE 3.27a
Example 3.19.

$$\{P_X\} = \{P_3\} = \{P_{AC}\}$$

Step 3: Identify the local coordinates
The combined local coordinates are indicated in Figure 3.27h.
Step 4: Develop $[T_{FA}]$ and $[T_{FX}]$.
The forces developed in the truss elements when unit loads are applied at the global coordinates are shown in Figures 3.27i and j. From these figures,

$$[T_{FA}] = \begin{bmatrix} 0.518 & -0.897 \\ 0 & 0 \\ -0.732 & -0.732 \end{bmatrix} \text{ and } [T_{FX}] = \begin{bmatrix} -0.897 \\ 1 \\ -0.732 \end{bmatrix}$$

Step 5: Develop $[f^i]$ and $[f]$

- Element stiffness matrix

$$[f^1] = [f^2] = [f^3] = \frac{l}{AE} = \frac{1}{50} \text{ mm/kN}$$

- Unassembled flexibility matrix

$$[f] = \frac{l}{AE} \begin{bmatrix} 1 & 0 & 0 \\ 0 & 1 & 0 \\ 0 & 0 & 1 \end{bmatrix}$$

Step 6: Find $[F_{XX}]$, $[F_{XA}]$, and $[F_{AA}]$

$$[F_{XX}] = \frac{l}{AE}[2.340]$$

$$[F_{XA}] = \frac{l}{AE} \begin{bmatrix} 0.071 & 1.34 \end{bmatrix}$$

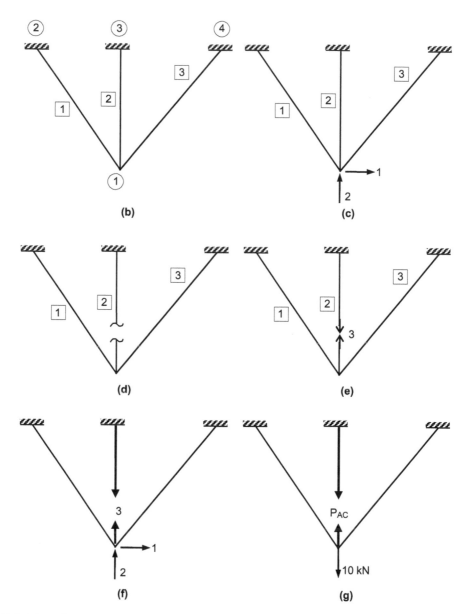

FIGURE 3.27b–g
Discretized structure, loads, and global coordinates. (b) Discretized structure. (c) Action coordinates. (d) Primary structure (DSI = 1). (e) Redundant coordinates. (f) Global coordinates. (g) Loads in primary structure.

$$[F_{AA}] = \frac{l}{AE} \begin{bmatrix} 0.804 & 0.0712 \\ 0.0712 & 1.340 \end{bmatrix}$$

Step 7: Determine $\{P_X\}$

$$\{D_X\} = \{0\}$$

$$\{P_X\} = P_3 = 5.727 \text{ kN}$$

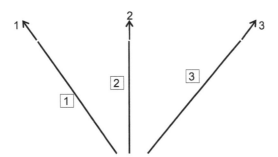

FIGURE 3.27h
Combined local coordinates.

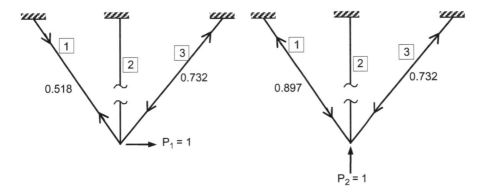

FIGURE 3.27i
Elements of $[T_{FA}]$.

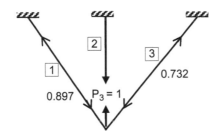

FIGURE 3.27j
Elements of $[T_{FX}]$.

Step 8: Determine the force in the truss elements $\{p\}$

$$\{p\} = [T_{FA}]\{P_A\} + [T_{FX}]\{P_X\}$$

$$\{p\} = \left\{ \begin{array}{c} p_1 \\ p_2 \\ p_3 \end{array} \right\} = \left\{ \begin{array}{c} 3.833 \\ 5.727 \\ 3.128 \end{array} \right\} kN$$

Step 9: Determine the nodal displacements $\{D_A\}$

$$\{D_A\} = [F_{AA}]\{P_A\} + [F_{AX}]\{P_X\}$$

$$\{D_A\} = \left\{ \begin{array}{c} D_1 \\ D_2 \end{array} \right\} = \frac{l}{AE} \left\{ \begin{array}{c} -0.304 \\ -5.727 \end{array} \right\} = \left\{ \begin{array}{c} -0.006 \\ -0.115 \end{array} \right\} \text{mm}$$

Example 3.20

Find the forces developed in the members of the truss shown in Figure 3.28a. Also find all the displacements at joints. AE is constant for all the members.

Solution

Step 1: Develop the discretized structure.
    The discretized structure is shown in Figure 3.28b.
Step 2: Identify the global coordinates and force vector. The DSI of the truss is 2 ($DSI_i = 1$, $DSI_e = 1$). The force in member BD (P_{BD}) and reaction at support C (C_X) are taken as the redundants.
    Since it is necessary to find all nodal displacements, all active coordinates are taken as action coordinates. The global coordinates, primary structure, and loads are shown in Figures 3.28c–g. The force vectors are

$$\{P_A\} = \left\{ \begin{array}{c} P_1 \\ P_2 \\ P_3 \\ P_4 \end{array} \right\} = \left\{ \begin{array}{c} -P \\ -2P \\ 0 \\ 0 \end{array} \right\}$$

$$\{P_X\} = \left\{ \begin{array}{c} P_5 \\ P_6 \end{array} \right\} = \left\{ \begin{array}{c} -C_X \\ P_{BD} \end{array} \right\}$$

Step 3: Identify the local coordinates
    The combined local coordinates are shown in Figure 3.28h.
Step 4: Develop $[T_{FA}]$ and $[T_{FX}]$
    The forces developed in the truss due to the application of unit loads at the global coordinates are shown in Figures 3.28i and j. From these figures,

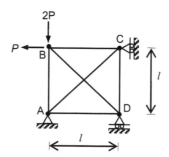

FIGURE 3.28a
Example 3.20.

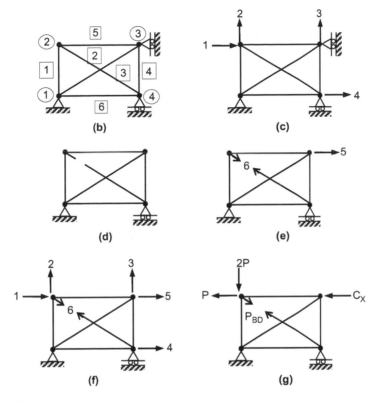

FIGURE 3.28b–g
Discretized structure, loads, and global coordinates. (b) Discretized structure. (c) Action coordinates. (d) Primary structure (DSI = 2, DSI$_i$ = 1, DSI$_e$ = 1). (e) Redundant coordinates. (f) Global coordinates. (g) Loads in primary structure.

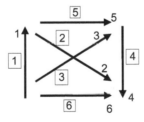

FIGURE 3.28h
Combined local coordinates.

$$[T_{FA}] = \begin{bmatrix} 0 & 1 & 0 & 0 \\ 0 & 0 & 0 & 0 \\ \sqrt{2} & 0 & 0 & 0 \\ -1 & 0 & 1 & 0 \\ -1 & 0 & 0 & 0 \\ 0 & 0 & 0 & 1 \end{bmatrix}$$

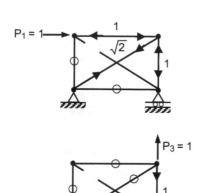
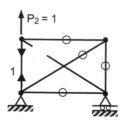
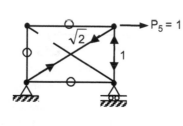
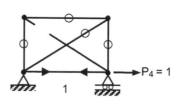

FIGURE 3.28i
Elements of $[T_{FA}]$.

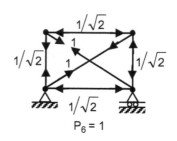

FIGURE 3.28j
Elements of $[T_{FX}]$.

$$[T_{FX}] = \begin{bmatrix} 0 & -1/\sqrt{2} \\ 0 & 1 \\ \sqrt{2} & 1 \\ -1 & -1/\sqrt{2} \\ 0 & -1/\sqrt{2} \\ 0 & -1/\sqrt{2} \end{bmatrix}$$

Step 5: Develop $[f^i]$ and $[f]$

- Element flexibility matrix

$$f^1 = f^4 = f^5 = f^6 = \frac{l}{AE}$$

$$f^2 = f^3 = \frac{\sqrt{2}l}{AE}$$

- Unassembled flexibility matrix

$$[f] = \frac{l}{AE} \begin{bmatrix} 1 & 0 & 0 & 0 & 0 & 0 \\ 0 & \sqrt{2} & 0 & 0 & 0 & 0 \\ 0 & 0 & \sqrt{2} & 0 & 0 & 0 \\ 0 & 0 & 0 & 1 & 0 & 0 \\ 0 & 0 & 0 & 0 & 1 & 0 \\ 0 & 0 & 0 & 0 & 0 & 1 \end{bmatrix}$$

Step 6: Find $[F_{XX}]$, $[F_{XA}]$, $[F_{AA}]$, and $[F_{AX}]$

$$[F_{AA}] = \frac{l}{AE} \begin{bmatrix} 4.828 & 0 & -1 & 0 \\ 0 & 1 & 0 & 0 \\ -1 & 0 & 1 & 0 \\ 0 & 0 & 0 & 1 \end{bmatrix}$$

$$[F_{XX}] = \frac{l}{AE} \begin{bmatrix} 3.828 & 2.707 \\ 2.707 & 4.828 \end{bmatrix}$$

$$[F_{XA}] = \frac{l}{AE} \begin{bmatrix} 3.828 & 0 & -1 & 0 \\ 3.414 & -0.707 & -0.707 & -0.707 \end{bmatrix}$$

$$[F_{AX}] = [F_{XA}]^T = \frac{l}{AE} \begin{bmatrix} 3.828 & 3.414 \\ 0 & -0.707 \\ -1 & -0.707 \\ 0 & -0.707 \end{bmatrix}$$

Step 7: Determine $\{P_X\}$

$$\{D_X\} = \{0\}$$

$$\{P_X\} = \begin{Bmatrix} 1.172P \\ -0.243P \end{Bmatrix}$$

Step 8: Find the force in truss elements $\{p\}$

$$\{p\} = [T_{FA}]\{P_A\} + [T_{FX}]\{P_X\}$$

$$\{p\} = \begin{Bmatrix} p_1 \\ p_2 \\ p_3 \\ p_4 \\ p_5 \\ p_6 \end{Bmatrix} = \begin{Bmatrix} -1.828P \\ -0.243P \\ 0 \\ 0 \\ 1.172P \\ 0.172P \end{Bmatrix}$$

Step 9: Determine the nodal displacements $\{D_A\}$

$$\{D_A\} = [F_{AA}]\{P_A\} + [F_{AX}]\{P_X\}$$

$$\{D_A\} = \begin{Bmatrix} D_1 \\ D_2 \\ D_3 \\ D_4 \end{Bmatrix} = \frac{Pl}{AE} \begin{Bmatrix} -1.172 \\ -1.828 \\ 0 \\ 0.172 \end{Bmatrix}$$

3.10 Analysis of Trusses Having Thermal Changes and Fabrication Errors

For the analysis of trusses having thermal changes and fabrication errors, the nodal forces (fixed end actions) required to prevent the movement of nodes of the truss element due to change in temperature or fabrication errors are found out. Next, equal and opposite forces (equivalent nodal loads) are applied at the nodes of the truss and the truss is analyzed for these loads. The actual force in the element of the truss is obtained by adding the element forces obtained for the above two load cases. This procedure is similar to analysis of structures subjected to element loads.

If a truss element of length l_i is made too long by an amount Δl_i (Figure 3.29a), then the nodal force (fixed end action) required to bring the element to its design length l_i is $-k_i\Delta l_i$, where k_i is the stiffness of the truss element ($k_i = A_iE_i/l_i$). The negative sign indicates that a compression force is applied to the element (Figure 3.29b).

In the case of analysis of truss due to increase in temperature ΔT, the increase in length is $\Delta l_i = l_i\alpha\Delta T$, where α is the coefficient of thermal expansion.

The forces shown in Figure 3.29b are with respect to the local axes. These forces are transformed to the direction of global axes to get the fixed end actions. After this, the procedures discussed in Sections 3.8 and 3.9 are used for the analysis of truss.

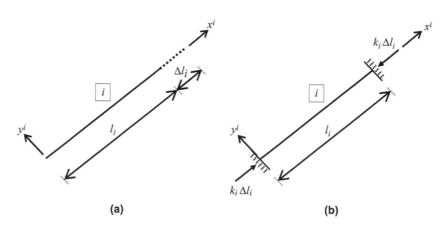

(a)   (b)

FIGURE 3.29
Truss element having fabrication error: (a) truss element and (b) fixed end actions (along local axes).

Example 3.21

Determine the forces and joint displacements of the truss shown in Figure 3.30a. The members AB and CB were made 5-mm long and 2-mm short before it was assembled. Take $AE/l = 100$ kN/mm.

Solution

Step 1: Develop the discretized structure.
   The nodes and elements are shown in Figure 3.30b.
Step 2: Determine the combined nodal loads
   The fixed end actions and equivalent nodal loads are indicated in Figures 3.30d–f. Since no nodal loads acts on the structure, the combined nodal loads are equal to the equivalent nodal loads.
Step 3: Identify the global coordinates and force vector.
   The structure is statically determinate. The global coordinates and the loads in the direction of global coordinates are shown in Figures 3.30g and h. The load vector is

$$\{P_{CA}\} = \begin{Bmatrix} P_{C1} \\ P_{C2} \end{Bmatrix} = \begin{Bmatrix} 358.579 \\ -141.421 \end{Bmatrix} \text{kN}$$

Step 4: Identify the local coordinates and $\{p_F\}$
   The combined local coordinates and element forces due to fixed end actions are shown in Figures 3.30i and j respectively.
   From Figures 3.30d, i and j

$$\{p_F\} = \begin{Bmatrix} -500 \\ 200 \end{Bmatrix} \text{kN}$$

Step 5: Develop $[T_{FA}]$
   The forces developed in the truss due to application of unit loads at different global coordinates are shown in Figure 3.30k.
   Hence,

$$[T_{FA}] = \begin{bmatrix} 1 & -1 \\ 0 & 1.414 \end{bmatrix}$$

Step 6: Develop $[f']$ and $[f]$

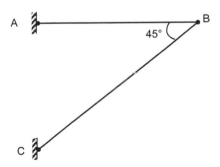

FIGURE 3.30a
Example 3.21.

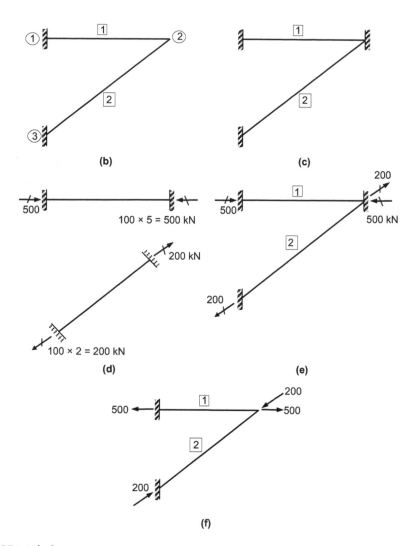

FIGURE 3.30b–f
Discretized structure, loads, and global coordinates. (b) Discretized structure. (c) Restrained structure. (d) Fixed end action of elements. (e) Fixed end action of structure. (f) Equivalent nodal loads.

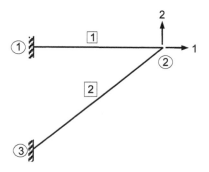

FIGURE 3.30g
Global coordinates.

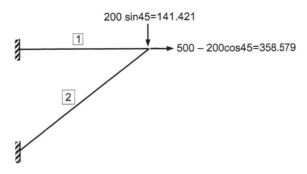

FIGURE 3.30h
Loads along the global coordinates.

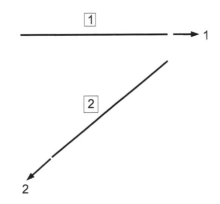

FIGURE 3.30i
Combined local coordinates.

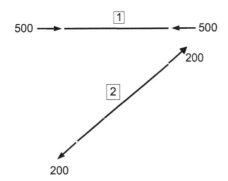

FIGURE 3.30j
Element loads due to fixed end actions.

- Element flexibility matrix

$$f^1 = f^2 = \frac{l}{AE} = \frac{1}{100} \text{ mm/kN}$$

- Unassembled flexibility matrix $[f]$

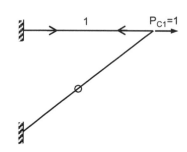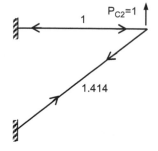

FIGURE 3.30k
Elements of $[T_{FA}]$.

$$[f] = \frac{1}{100} \begin{bmatrix} 1 & 0 \\ 0 & 1 \end{bmatrix}$$

Step 7: Find $[F_{AA}]$

$$[F_{AA}] = [T_{FA}]^T [f][T_{FA}]$$

$$= \frac{1}{100} \begin{bmatrix} 1 & -1 \\ -1 & 3 \end{bmatrix}$$

Step 8: Determine the nodal displacements $\{D_A\}$

$$\{D_A\} = [F_{AA}]\{P_{CA}\}$$

$$\{D_A\} = \left\{ \begin{array}{c} D_1 \\ D_2 \end{array} \right\} = \left\{ \begin{array}{c} 5 \\ -7.828 \end{array} \right\} \text{mm}$$

Step 9: Determine the force in elements $\{p\}$

$$\{p\} = \{p_F\} + [T_{FA}]\{P_{CA}\} + [T_{FX}]\{P_{CX}\}$$

$$\{p\} = \left\{ \begin{array}{c} p_1 \\ p_2 \end{array} \right\} = \left\{ \begin{array}{c} 0 \\ 0 \end{array} \right\}$$

Comments

From the above example, it can be seen that fabrication errors or change in temperature of the elements does not lead to development of element forces in a statically determinate structure. Similarly, it can be shown that settlement of supports in a statically determinate structure will not develop element forces.

Example 3.22

Determine the forces developed and joint displacements of the truss shown in Figure 3.31a if the members AB, AC, and AE were made 5-mm short, 1-mm long, and 2-mm short before the fabrication of truss. Take $AE/l = 100$ kN/mm.

Solution

Step 1: Develop the discretized structure.

The nodes and elements are shown in Figure 3.31b.

Step 2: Determine the combined nodal loads.

The restrained structure, fixed end actions, and combined nodal loads are shown in Figures 3.31c–f.

Step 3: Identify the global coordinates and force vector.

The DSI of the truss is 2 ($DSI_i = 2$). The forces in members AD and AE (P_{AD} and P_{AE}) are chosen as the redundant forces.

The primary structure, global coordinates, and loads are shown in Figures 3.31g–l. The force vectors are

$$\{P_{CA}\} = \left\{ \begin{array}{c} P_{C1} \\ P_{C2} \end{array} \right\} = \left\{ \begin{array}{c} -180.34 \\ 353.55 \end{array} \right\} \text{kN}$$

$$\{P_{CX}\} = \left\{ \begin{array}{c} P_{C3} \\ P_{C4} \end{array} \right\} = \left\{ \begin{array}{c} P_{AD} \\ P_{AE} \end{array} \right\}$$

Step 4: Identify the local coordinates and $\{p_F\}$

The combined local coordinates are shown in Figure 3.31m. From Figures 3.31d and n,

$$\{p_F\} = \left\{ \begin{array}{c} 500 \\ -100 \\ 0 \\ 200 \end{array} \right\} \text{kN}$$

Step 5: Develop $[T_{FA}]$ and $[T_{FX}]$

Figures 3.31o and p,

$$[T_{FA}] = \left[\begin{array}{cc} 1.414 & 0 \\ -1 & -1 \\ 0 & 0 \\ 0 & 0 \end{array} \right] \text{ and } [T_{FX}] = \left[\begin{array}{cc} 0.707 & 1.225 \\ -1.366 & -1.366 \\ 1 & 0 \\ 0 & 1 \end{array} \right]$$

Step 6: Develop $[f^i]$ and $[f]$

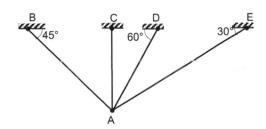

FIGURE 3.31a
Example 3.22.

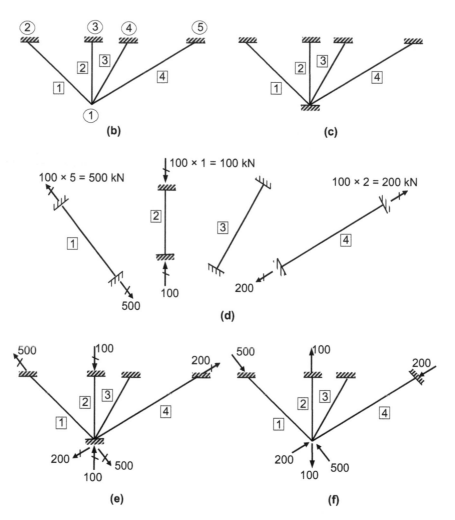

FIGURE 3.31b–f
Discretized structure, loads, and global coordinates. (b) Discretized structure. (c) Restrained
structure. (d) Fixed end action of elements. (e) Fixed end action of structure. (f) Equivalent nodal
loads.

- Element flexibility matrix

$$f^1 = f^2 = f^3 = f^4 = \frac{l}{AE} = \frac{1}{100} \text{ mm/kN}$$

- Unassembled flexibility matrix

$$[f] = \frac{1}{100} \begin{bmatrix} 1 & 0 & 0 & 0 \\ 0 & 1 & 0 & 0 \\ 0 & 0 & 1 & 0 \\ 0 & 0 & 0 & 1 \end{bmatrix}$$

Step 7: Find $[F_{XX}]$, $[F_{XA}]$, and $[F_{AA}]$

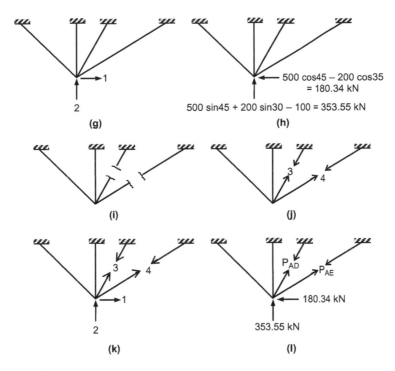

FIGURE 3.31g–l
Global coordinates and loads. (g) Action coordinates. (h) Loads along action coordinates. (i) Primary structure. (j) Redundant coordinates. (k) Global coordinates. (l) Loads on primary structure.

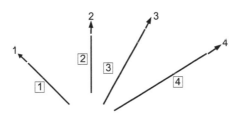

FIGURE 3.31m
Combined local coordinates.

$$[F_{XX}] = \begin{bmatrix} 0.0337 & 0.0273 \\ 0.0273 & 0.0437 \end{bmatrix}$$

$$[F_{XA}] = \begin{bmatrix} 0.0237 & 0.0137 \\ 0.0310 & 0.0137 \end{bmatrix}$$

$$[F_{AA}] = \begin{bmatrix} 0.03 & 0.01 \\ 0.01 & 0.01 \end{bmatrix}$$

Step 8: Determine $\{P_{CX}\}$

$$\{D_A\} = \{0\}$$

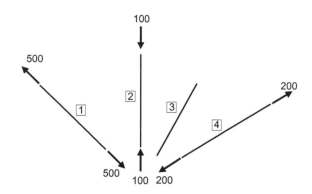

FIGURE 3.31n
Element forces due to fixed end actions.

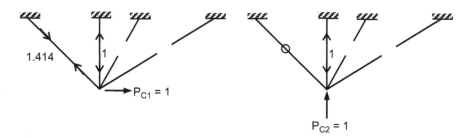

FIGURE 3.31o
Elements of $[T_{FA}]$.

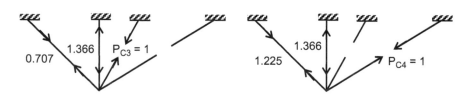

FIGURE 3.31p
Elements of $[T_{FX}]$.

$$\{P_{CX}\} = \left\{ \begin{array}{c} P_{C3} \\ P_{C4} \end{array} \right\} = \left\{ \begin{array}{c} -62.62 \\ 56.53 \end{array} \right\}$$

Step 9: Determine the force in elements $\{p\}$

$$\{p\} = \{p_F\} + [T_{FA}]\{P_{CA}\} + [T_{FX}]\{P_{CX}\}$$

$$\{p\} = \left\{ \begin{array}{c} p_1 \\ p_2 \\ p_3 \\ p_4 \end{array} \right\} = \left\{ \begin{array}{c} 269.98 \\ -264.9 \\ -62.62 \\ 256.53 \end{array} \right\} \text{kN}$$

Step 10: Determine the nodal displacements $\{D_A\}$

$$\{D_A\} = [F_{AA}]\{P_{CA}\} + [F_{AX}]\{P_{CX}\}$$

$$\{D_A\} = \left\{ \begin{array}{c} D_1 \\ D_2 \end{array} \right\} = \left\{ \begin{array}{c} -1.604 \\ 1.649 \end{array} \right\} \text{mm}$$

Problem

3.1 Analyze the following structures (Figure 3.32) using the flexibility method.

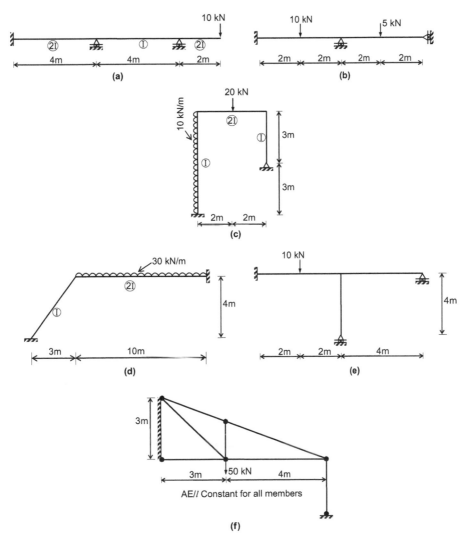

FIGURE 3.32
Problem 3.1.

4

Stiffness Method

4.1 Introduction

In the previous chapter, the flexibility method was used for the analysis of structures. It was seen that, in flexibility approach, different procedures are required for the analysis of statically determinate and statically indeterminate structures. Moreover, by choosing different redundant forces, there are many ways by which a statically indeterminate structure can be analyzed. In other words, the choice of unknowns (i.e., redundant forces) is not unique in the flexibility method. All these aspects make the flexibility method not suitable for computer programming.

The above limitations of the flexibility method can be avoided by using the stiffness method. In this method, the active DOF of the structure are the basic unknowns, which are unique for a given structure. Furthermore, the procedure for analysis remains the same for all type of structures. Hence, the stiffness method is more suitable for automatic analysis of structures.

In the stiffness method, first of all, the compatibility conditions are used to relate the end displacements of the elements with the nodal displacements of the structure. Then the force–displacement relations using stiffness coefficients are established between the end actions and displacements of the elements, and between the nodal forces and displacements of the structure. Finally, from the equations of equilibrium using the FBD of the nodes, the unknown nodal displacements (active DOF) are determined. Using this, the support reactions, element forces, and element displacements are found out. The orders in which the three basic equations are used in the stiffness method of analysis are shown in Figure 4.1.

In this chapter, the equations necessary for analysis using the stiffness method are derived and the procedure for analysis are explained by considering beams, plane frames, and plane trusses.

4.2 Coordinates for Displacements and Forces

In the stiffness method, the displacements are treated as independent variables. Hence in this method, coordinates indicate displacements. Once the coordinates are chosen, then they also indicate the direction of corresponding forces.

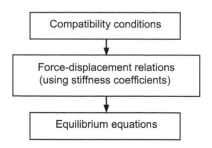

FIGURE 4.1
Basic equations used in the stiffness method.

4.2.1 Global Coordinates: Active and Restrained Coordinates

Global coordinates indicate the nodal displacements of the structure, i.e., they define the DOF of the structure. As discussed in Section 1.10, there are two types of DOF, namely active and restrained. The coordinates corresponding to active DOF and restrained DOF are known as *active coordinates* and *restrained coordinates*. Thus in the stiffness method, the nodal displacement vector $\{D\}$ is partitioned as

$$\{D\} = \left\{ \begin{array}{c} \{D_A\} \\ \hline \{D_R\} \end{array} \right\} \tag{4.1}$$

where $\{D_A\}$ is the *active displacement vector* and is the primary unknown in the stiffness method. The number of elements in $\{D_A\}$ is equal to the DKI of the structure. $\{D_R\}$ is the *restrained displacement vector* and is the known vector.

The force vector $\{P\}$ in the stiffness method is partitioned as

$$\{P\} = \left\{ \begin{array}{c} \{P_A\} \\ \hline \{P_R\} \end{array} \right\} \tag{4.2}$$

where $\{P_A\}$ and $\{P_R\}$ are the force vectors corresponding to $\{D_A\}$ and $\{D_R\}$, respectively. The active coordinates also indicate the locations where the loads act on a structure. Hence, $\{P_A\}$ is known as the *load vector*, which is a known quantity. The forces along the restrained coordinates are the reactions, which are not known. Hence, $\{P_R\}$ is called the *reaction vector*.

In order to partition the global coordinates into active and restrained coordinates, while numbering the global coordinates, the active DOF are numbered before the restrained DOF. The global coordinates for typical framed structures are shown in Figure 4.2.

$$\{P_A\} = \left\{ \begin{array}{c} P_1 \\ P_2 \\ P_3 \\ P_4 \\ P_5 \\ P_6 \\ P_7 \end{array} \right\} = \left\{ \begin{array}{c} 0 \\ -P_c \\ 0 \\ 0 \\ 0 \\ P_{bx} \\ -P_{by} \end{array} \right\} \qquad \{P_R\} = \left\{ \begin{array}{c} P_8 \\ P_9 \\ P_{10} \end{array} \right\} = \left\{ \begin{array}{c} a_x \\ a_y \\ e_y \end{array} \right\}$$

Load vector (Structure 1)   Reaction vector (Structure 1)

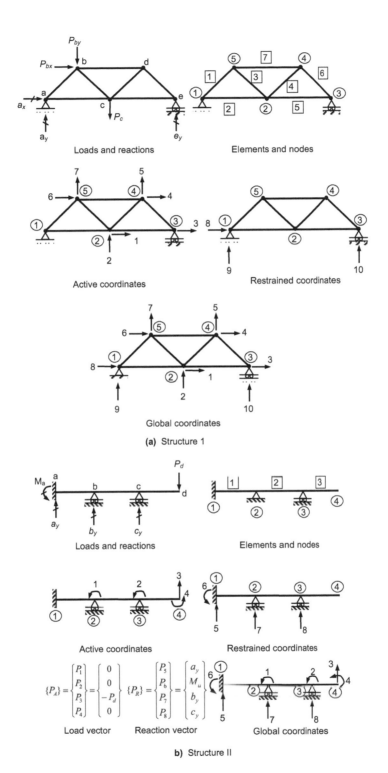

FIGURE 4.2
Global coordinates—Stiffness method.

(Continued)

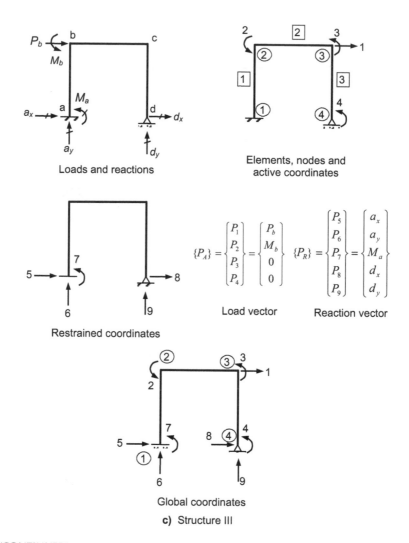

Loads and reactions

Elements, nodes and active coordinates

Restrained coordinates

$$\{P_A\} = \begin{Bmatrix} P_1 \\ P_2 \\ P_3 \\ P_4 \end{Bmatrix} = \begin{Bmatrix} P_b \\ M_b \\ 0 \\ 0 \end{Bmatrix} \quad \{P_R\} = \begin{Bmatrix} P_5 \\ P_6 \\ P_7 \\ P_8 \\ P_9 \end{Bmatrix} = \begin{Bmatrix} a_x \\ a_y \\ M_a \\ d_x \\ d_y \end{Bmatrix}$$

Load vector   **Reaction vector**

Global coordinates

c) Structure III

FIGURE 4.2 (CONTINUED)
Global coordinates—Stiffness method.

Structure I (Figure 4.2a) is a plane truss. Hence, there are two DOF at a node. The active and restrained coordinates and the force vectors are shown in the figure. The global coordinates of the beam (Structure II) are shown in Figure 4.2b. For a beam, the number of DOF at a node is again 2. The coordinates of a plane frame (Structure III) are shown in Figure 4.2c. The DOF per node is 3. The global coordinates neglecting axial deformations are indicated in the figure.

Comments

The reader should understand the difference between the action coordinates used in the flexibility method and the active coordinates introduced in the stiffness method. Both these coordinates are along the free DOF. However, all the free DOF need not be chosen as action coordinates. The loads can be applied only along the free DOF. Hence, in the flexibility method, the free DOF at which the loads are applied and the coordinates at which nodal displacements are to be found out are chosen as action coordinates. Whereas in the stiffness method, the free nodal DOF are the basic unknowns. The number of free DOF is the DKI of the structure. Hence, all the free DOF must be chosen as the active coordinates so that the unknown nodal displacements are determined. Thus in a general case, the number of action coordinates will be less than the number of active coordinates. But when it becomes necessary to choose all the free DOF as action coordinates, then in such cases, the action and active coordinates will be the same for the structure.

4.2.2 Local Coordinates

4.2.2.1 Elements, Coordinates, and Stiffness Matrix

Figure 4.3 shows the end actions and the corresponding end displacements of a plane frame element, beam element, and truss element. The stiffness matrices $[k^i]$ of these elements are derived in Example 2.2. The force–displacement relations of these elements are

$$\{p^i\} = [k^i]\{d^i\}$$

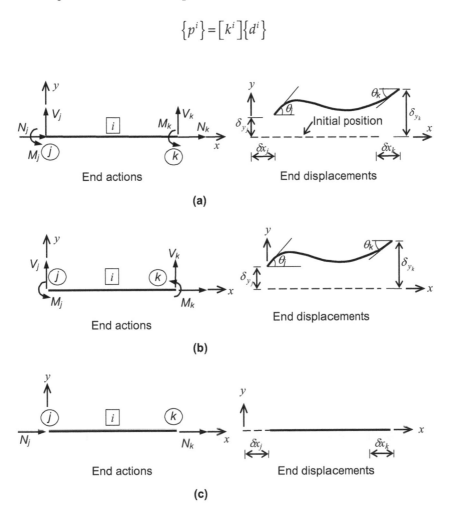

FIGURE 4.3
General end actions and end displacements in elements: (a) plane frame element, (b) beam element, and (c) plane truss element.

- Plane frame element

$$
\begin{Bmatrix} N_j \\ V_j \\ M_j \\ N_k \\ V_k \\ M_k \end{Bmatrix} =
\begin{bmatrix}
AE/l & 0 & 0 & -AE/l & 0 & 0 \\
0 & 12EI/l^3 & 6EI/l^2 & 0 & -12EI/l^3 & 6EI/l^2 \\
0 & 6EI/l^2 & 4EI/l & 0 & -6EI/l^2 & 2EI/l \\
-AE/l & 0 & 0 & AE/l & 0 & 0 \\
0 & -12EI/l^3 & -6EI/l^2 & 0 & 12EI/l^3 & -6EI/l^2 \\
0 & 6EI/l^2 & 2EI/l & 0 & -6EI/l^2 & 4EI/l
\end{bmatrix}
\begin{Bmatrix} \delta_{xj} \\ \delta_{yj} \\ \theta_j \\ \delta_{xk} \\ \delta_{yk} \\ \theta_k \end{Bmatrix}
$$

(4.3a)

- Beam element

$$
\begin{Bmatrix} V_j \\ M_j \\ V_k \\ M_k \end{Bmatrix} =
\begin{bmatrix}
12EI/l^3 & 6EI/l^2 & -12EI/l^3 & 6EI/l^2 \\
6EI/l^2 & 4EI/l & -6EI/l^2 & 2EI/l \\
-12EI/l^3 & -6EI/l^2 & 12EI/l^3 & -6EI/l^2 \\
6EI/l^2 & 2EI/l & -6EI/l^2 & 4EI/l
\end{bmatrix}
\begin{Bmatrix} \delta_{yj} \\ \theta_j \\ \delta_{yk} \\ \theta_k \end{Bmatrix}
$$

(4.3b)

- Truss element

$$
\begin{Bmatrix} N_j \\ N_k \end{Bmatrix} =
\begin{bmatrix}
AE/l & -AE/l \\
-AE/l & AE/l
\end{bmatrix}
\begin{Bmatrix} \delta_{xj} \\ \delta_{xk} \end{Bmatrix}
$$

(4.3c)

4.2.2.2 Basic End Actions, Displacements, and Stiffness Matrix

In Section 3.2.2, it was seen that for a plane frame element, out of the six end actions, only three actions are independent or basic. N_k, M_j, and M_k were chosen as the basic end actions. From Figure 4.4, it is observed that the axial deformation caused due to displacements δ_{xj} and δ_{xk} is adequately represented by ε. Here, ε is the elongation of the element and is defined as

$$
\varepsilon = \delta_{xk} - \delta_{xj}
$$

(4.4a)

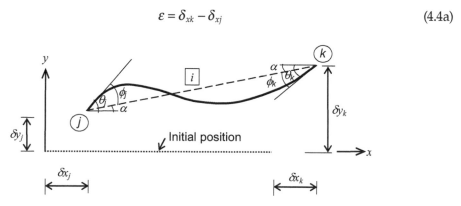

FIGURE 4.4
General and basic end displacements.

Similarly, the flexural deformations of the element are sufficiently described by the rotations ϕ_j and ϕ_k are measured with respect to the chord connecting the ends of the element. However, the rotations θ_j and θ_k are measured with respect to the local x-axis. From the figure,

$$\theta_j = \phi_j + \alpha \tag{4.4b}$$

$$\theta_k = \phi_k + \alpha \tag{4.4c}$$

where

$$\alpha = \frac{\delta_{yk} - \delta_{yj}}{l} \tag{4.4d}$$

ε, ϕ_j, and ϕ_k are the basic end displacements and N_k, M_j, and N_k are the corresponding end actions. The basic end actions and displacements of different types of elements are shown in Figure 4.5.

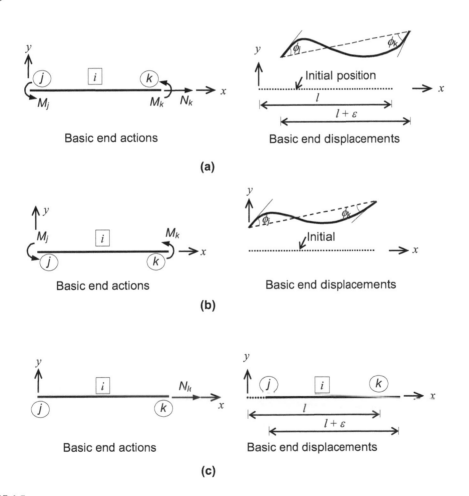

FIGURE 4.5
Basic end actions and displacements of elements: (a) plane frame element, (b) beam element, and (c) plane truss element.

The relations connecting the basic end actions and displacements can be derived using Eq. 4.3. For a plane frame element, from Eq. 4.3a,

$$N_k = -\frac{AE}{l}\delta_{xj} + \frac{AE}{l}\delta_{xk} = \frac{AE}{l}(\delta_{xk} - \delta_{xj}) = \frac{AE}{l}\varepsilon$$

$$M_j = \frac{6EI}{l^2}\delta_{yj} + \frac{4EI}{l}\theta_j - \frac{6EI}{l^2}\delta_{yk} + \frac{2EI}{l}\theta_k \qquad\qquad \text{(a)}$$

$$= -\frac{6EI}{l}\frac{(\delta_{yk} - \delta_{yj})}{l} + \frac{4EI}{l}\theta_j + \frac{2EI}{l}\theta_k$$

From Eq. 4.4,

$$\frac{(\delta_{yk} - \delta_{yj})}{l} = \alpha, \quad \theta_j = \theta_j + \alpha \text{ and } \theta_k = \phi_k + \alpha$$

Substituting in the equation for M_j gives

$$M_j = -\frac{6EI}{l}\alpha + \frac{4EI}{l}(\theta_j + \alpha) + \frac{2EI}{l}(\theta_k + \alpha)$$

$$= \frac{4EI}{l}\phi_j + \frac{2EI}{l}\phi_k \qquad\qquad\qquad \text{(b)}$$

Similarly,

$$M_k = \frac{6EI}{l^2}\delta_{yj} + \frac{2EI}{l}\theta_j - \frac{6EI}{l^2}\delta_{yk} + \frac{4EI}{l}\theta_k$$

$$= -\frac{6EI}{l^2}(\delta_{yk} - \delta_{yj}) + \frac{2EI}{l}\theta_j + \frac{4EI}{l}\theta_k \qquad\qquad \text{(c)}$$

$$= \frac{2EI}{l}\phi_j + \frac{4EI}{l}\phi_k$$

Equations (a) to (c) in matrix form give

$$\begin{Bmatrix} N_k \\ M_j \\ M_k \end{Bmatrix} = \begin{bmatrix} AE/l & 0 & 0 \\ 0 & 4EI/l & 2EI/l \\ 0 & 2EI/l & 4EI/l \end{bmatrix} \begin{Bmatrix} \varepsilon \\ \phi_j \\ \phi_k \end{Bmatrix} \qquad\qquad \text{(4.5a)}$$

or

$$\{p^i\} = [k^i]\{d^i\}$$

For a beam element,

$$\begin{Bmatrix} M_j \\ M_k \end{Bmatrix} = \begin{bmatrix} 4EI/l & 2EI/l \\ 2EI/l & 4EI/l \end{bmatrix} \begin{Bmatrix} \phi_j \\ \phi_k \end{Bmatrix} \qquad\qquad \text{(4.5b)}$$

and for a plane truss element,

$$N_k = \frac{AE}{l}\varepsilon \qquad (4.5c)$$

In the above equations, $[k^i]$ which relates the basic end actions with the basic end displacements is known as the *basic element stiffness matrix*. Compared to the element stiffness matrix given in Eq. 4.3, the size of the element stiffness matrix is reduced when basic end actions and displacements are used. Hence, $[k^i]$ is Eq. 4.5 is also known as the *reduced element stiffness matrix*.

The number of coordinates for the elements reduces from six to three for a plane frame element, four to two in beam element, and two to one for a truss element. In other words, the computational effort is greatly reduced if Eq. 4.5 is used. Hence in this chapter, in order to reduce the hand computational work, only the basic end actions and displacements are considered in analysis. Thus, the local coordinates and combined local coordinates discussed in the flexibility method (see Figure 3.5) are also applicable in the stiffness method of analysis.

4.3 Compatibility Conditions and Displacement Transformation Matrix

The compatibility condition used in the stiffness method is that the geometry of the deformation must be such that the deformed elements of the structure should fit together at nodes. For this to happen, the end displacements of the elements (i.e., displacements in local coordinates) should be related to the nodal displacements of the structure (i.e., displacements in global coordinates). Hence, each element displacement is written in terms of all the nodal displacements as

$$
\begin{aligned}
d_1 &= T_{D11}D_1 + T_{D12}D_2 + \cdots + T_{D1n}D_n \\
d_2 &= T_{D21}D_1 + T_{D22}D_2 + \cdots + T_{D2n}D_n \\
&\cdots\cdots\cdots\cdots\cdots\cdots\cdots\cdots\cdots\cdots\cdots\cdots \qquad (a) \\
&\cdots\cdots\cdots\cdots\cdots\cdots\cdots\cdots\cdots\cdots\cdots\cdots \\
d_m &= T_{Dm1}D_1 + T_{Dm2}D_2 + \cdots + T_{Dmn}D_n
\end{aligned}
$$

or

$$\{d\} = [T_D]\{D\} \qquad (4.6)$$

where

$\{d\}$ is the combined element displacement vector (see Section 3.2.2), $\{D\}$ is the displacement vector and

$$
[T_D] = \begin{bmatrix}
T_{D11} & T_{D12} & \cdots & T_{D1n} \\
T_{D21} & T_{D22} & \cdots & T_{D2n} \\
\cdots & \cdots & \cdots & \cdots \\
T_{Dm1} & T_{Dm2} & \cdots & T_{Dmn}
\end{bmatrix}
$$

is the *displacement transformation matrix*.

$$
\begin{array}{cccccc}
D_1{=}1 & D_2{=}1 & & D_j{=}1 & & D_n{=}1 \\
\end{array}
$$

$$
[T_D] = \begin{bmatrix}
T_{D11} & T_{D12} & \cdots & T_{D1j} & \cdots & T_{D1n} \\
T_{D21} & T_{D22} & \cdots & T_{D2j} & \cdots & T_{D2n} \\
T_{D31} & T_{D32} & \cdots & T_{D3j} & \cdots & T_{D3n} \\
\cdots & \cdots & \cdots & \cdots & \cdots & \cdots \\
T_{Dm1} & T_{Dm2} & \cdots & T_{Dmj} & \cdots & T_{Dmn}
\end{bmatrix}
$$

FIGURE 4.6
Development of $[T_D]$.

$[T_D]$ is a transformation matrix that transforms the displacements from the global coordinates to local coordinates. The displacement transformation influence coefficient T_{Dij} is defined as the displacement developed in the element corresponding to i^{th} combined local coordinate due to a unit displacement at the jth global coordinate, while the displacements at all other global coordinates are zero. Similar to the force transformation matrix, the displacement transformation matrix is developed column-wise. To get the elements of jth column of $[T_D]$, a unit displacement is applied at the jth global coordinate and the displacements at other global coordinates are made zero by introducing appropriate support conditions.* From the deformed shape of the structure caused by these nodal displacements, the element displacements are determined. The combined element displacement vector for this case will be the elements of jth column of $[T_D]$ (Figure 4.6).

The displacement transformation matrix $[T_D]$ is partitioned into two submatrices $[T_{DA}]$ and $[T_{DR}]$ to indicate the separate influences of $\{D_A\}$ and $\{D_R\}$. Hence, Eq. 4.6 can be written as

$$
[T_D] = \left[\ \ [T_{DA}]\ \vdots\ [T_{DR}]\ \ \right]
$$

$$
\Rightarrow\ [d] = \left[\ \ [T_{DA}]\ \vdots\ [T_{DR}]\ \ \right] \left\{ \begin{array}{c} \{D_A\} \\ \hline \{D_R\} \end{array} \right\} \tag{4.7a}
$$

or

$$
[d] = [T_{DA}]\{D_A\} + [T_{DR}]\{D_R\}] \tag{4.7b}
$$

Example 4.1

Develop the displacement transformation matrix for the structures shown in Figure 4.7a.

Solution

Structure I

The discretized structure, global coordinates, local coordinates, and restrained structure are shown in Figure 4.7b.

---

* Alternatively, this is equivalent to apply a unit displacement at the jth global coordinate of the restrained structure.

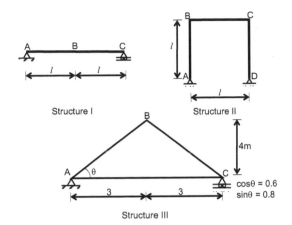

FIGURE 4.7a
Example 4.1.

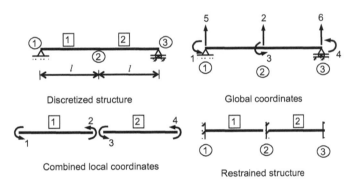

FIGURE 4.7b
Discretized structure and coordinates—Structure I.

- Elements of the first column of $[T_D]$

  To get the elements of the first column of $[T_D]$, a unit displacement is applied at the global coordinate 1 ($D_1 = 1$) and the displacements at other coordinates are made zero. Appropriate support conditions are introduced so as to get the required deformed condition for the structure. This is same as applying a unit displacement at the first global coordinate in the restrained structure (Figure 4.7c).

  The element deformations are obtained from the deformed shape of the structure. In the deformed structures, the angle is measured from the chord to the tangent. The combined local coordinates (Figure 4.7b) indicate the positive direction of element deformations. From Figure 4.7c ($D_1 = 1$),

  $$d_1 = T_{D11} = 1, \quad d_2 = T_{D21} = 0, \quad d_3 = T_{D31} = 0, \quad d_4 = T_{D41} = 0$$

- Elements of the second column of $[T_D]$

  From Figure 4.7c ($D_2 = 1$), it can be seen that d_1 and d_2 are in clockwise direction. Hence, they are negative quantities. d_3 and d_4 are in the counter clockwise direction. Thus, they are positive.

  $$d_1 = T_{D12} = -1/l, \quad d_2 = T_{D22} = -1/l, \quad d_3 = T_{D32} = 1/l, \quad d_4 = T_{D42} = 1/l,$$

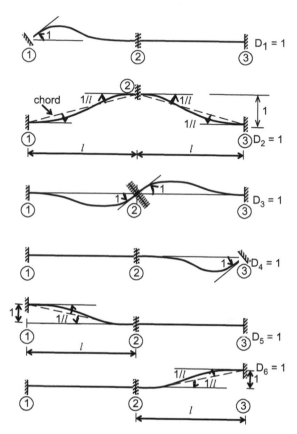

FIGURE 4.7c
Elements of $[T_D]$—Structure I.

- Elements of the third column of $[T_D]$
  From Figure 4.7c ($D_3 = 1$),

$$d_1 = T_{D13} = 0, \quad d_2 = T_{D23} = 1, \quad d_3 = T_{D33} = 1, \quad d_4 = T_{D43} = 0$$

- Elements of the fourth column of $[T_D]$
  From Figure 4.7c ($D_4 = 1$),

$$d_1 = T_{D14} = 0, \quad d_2 = T_{D24} = 0, \quad d_3 = T_{D34} = 0, \quad d_4 = T_{D44} = 1$$

- Elements of the fifth column of $[T_D]$
  From Figure 4.7c ($D_5 = 1$),

$$d_1 = T_{D15} = 1/l, \quad d_2 = T_{D25} = 1/l, \quad d_3 = T_{D35} = 0, \quad d_4 = T_{D45} = 0$$

- Elements of the sixth column of $[T_D]$
  From Figure 4.7c ($D_6 = 1$),

$$d_1 = T_{D16} = 0, \quad d_2 = T_{D26} = 0, \quad d_3 = T_{D36} = -1/l, \quad d_4 = T_{D46} = -1/l$$

Displacement transformation matrix $[T_D]$

- $[T_{DA}]$

    In this case, the active coordinates are 1 to 4. Hence,

$$[T_{DA}] = \begin{array}{cccc} \scriptstyle D_1=1 & \scriptstyle D_2=1 & \scriptstyle D_3=1 & \scriptstyle D_4=1 \\ \left[\begin{array}{c|c|c|c} 1 & -1/l & 0 & 0 \\ 0 & -1/l & 1 & 0 \\ 0 & 1/l & 1 & 0 \\ 0 & 1/l & 0 & 1 \end{array}\right] \end{array}$$

- $[T_{DR}]$

    The global coordinates 5 and 6 are restrained coordinates. Hence,

$$[T_{DR}] = \begin{array}{cc} \scriptstyle D_5=1 & \scriptstyle D_6=1 \\ \left[\begin{array}{c|c} 1/l & 0 \\ 1/l & 0 \\ 0 & -1/l \\ 0 & -1/l \end{array}\right] \end{array}$$

- $[T_D]$

$$[T_D] = \left[\begin{array}{c|c} [T_{DA}] & [T_{DR}] \end{array} \right]$$

$$= \begin{array}{cccccc} \scriptstyle D_1=1 & \scriptstyle D_2=1 & \scriptstyle D_3=1 & \scriptstyle D_4=1 & \scriptstyle D_5=1 & \scriptstyle D_6=1 \\ \left[\begin{array}{c|c|c|c|c|c} 1 & -1/l & 0 & 0 & 1/l & 0 \\ 0 & -1/l & 1 & 0 & 1/l & 0 \\ 0 & 1/l & 1 & 0 & 0 & 1/l \\ 0 & 1/l & 0 & 1 & 0 & -1/l \end{array}\right] \end{array}$$

Structure II

The discretized structure, global coordinates, local coordinates, and restrained structure are shown in Figure 4.7d.

The deformed shape of the structure due to the application of unit displacements at the global coordinates is shown in Figure 4.7e.

- Element deformations due to $D_1 = 1$

$$d_1 = 1/l, \quad d_2 = 1/l, \quad d_3 = 0, \quad d_4 = 0, \quad d_5 = 1/l, \quad d_6 = 1/l$$

- Element deformations due to $D_2 = 1$

$$d_1 = 0, \quad d_2 = 1, \quad d_3 = 1, \quad d_4 = 0, \quad d_5 = 0, \quad d_6 = 0$$

- Element deformations due to $D_3 = 1$

$$d_1 = 0, \quad d_2 = 0, \quad d_3 = 0, \quad d_4 = 1, \quad d_5 = 1, \quad d_6 = 0$$

- Element deformations due to $D_4 = 1$

$$d_1 = 0, \quad d_2 = 0, \quad d_3 = 0, \quad d_4 = 0, \quad d_5 = 0, \quad d_6 = 1$$

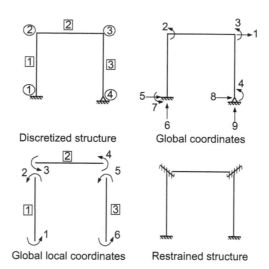

FIGURE 4.7d
Discretized structure and coordinates—Structure II.

- Element deformations due to $D_5 = 1$

$$d_1 = -1/l, \quad d_2 = -1/l, \quad d_3 = 0, \quad d_4 = 0, \quad d_5 = 0, \quad d_6 = 0$$

- Element deformations due to $D_6 = 1$

$$d_1 = 0, \quad d_2 = 0, \quad d_3 = 1/l, \quad d_4 = 1/l, \quad d_5 = 0, \quad d_6 = 0$$

- Element deformations due to $D_7 = 1$

$$d_1 = 1, \quad d_2 = 0, \quad d_3 = 0, \quad d_4 = 0, \quad d_5 = 0, \quad d_6 = 0$$

- Element deformations due to $D_8 = 1$

$$d_1 = 0, \quad d_2 = 0, \quad d_3 = 0, \quad d_4 = 0, \quad d_5 = -1/l, \quad d_6 = -1/l$$

- Element deformations due to $D_9 = 1$

$$d_1 = 0, \quad d_2 = 0, \quad d_3 = -1/l, \quad d_4 = -1/l, \quad d_5 = 0, \quad d_6 = 0$$

Displacement transformation matrix
The global coordinates 1 to 4 are active and the remaining are restrained

- $[T_{DA}]$

$$[T_{DA}] = \begin{array}{c} \begin{matrix} D_1=1 & D_2=1 & D_3=1 & D_4=1 \end{matrix} \\ \begin{bmatrix} 1/l & 0 & 0 & 0 \\ 1/l & 1 & 0 & 0 \\ 0 & 1 & 0 & 0 \\ 0 & 0 & 1 & 0 \\ 1/l & 0 & 1 & 0 \\ 1/l & 0 & 0 & 1 \end{bmatrix} \end{array}$$

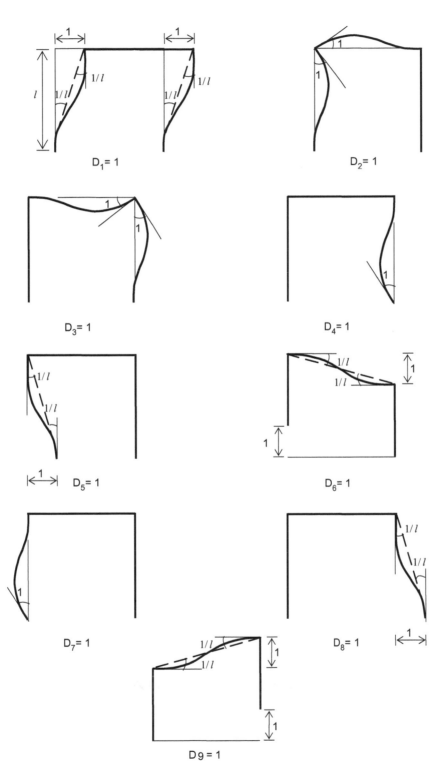

FIGURE 4.7e
Deformed shape of Structure II.

- $[T_{DR}]$

$$[T_{DR}]=\begin{bmatrix} \overset{D_5=1}{-1/l} & \overset{D_6=1}{0} & \overset{D_7=1}{1} & \overset{D_8=1}{0} & \overset{D_9=1}{0} \\ -1/l & 0 & 0 & 0 & 0 \\ 0 & 1/l & 0 & 0 & -1/l \\ 0 & 1/l & 0 & 0 & -1/l \\ 0 & 0 & 0 & -1/l & 0 \\ 0 & 0 & 0 & -1/l & 0 \end{bmatrix}$$

- $[T_D]$

$$[T_D]=\begin{bmatrix} [T_{DA}] & \vdots & [T_{DR}] \end{bmatrix}$$

Structure III

The discretized structure, coordinates, and restrained structure are shown in Figure 4.7f. The deformed shape of the structure and the element deformations are shown in Figure 4.7g.

- Element deformations due to $D_1=1$

$$d_1 = \cos\theta = 0.6, \quad d_2 = -\cos\theta = -0.6, \quad d_3 = 0$$

A positive value of element deformation indicates that it is along the positive direction of local coordinates. In the case of truss element, it also indicates elongation of the element. Similarly, a negative element deformation indicates shortening of the truss element.

- Element deformations due to $D_2=1$

$$d_1 = \sin\theta = 0.8, \quad d_2 = \sin\theta = 0.8, \quad d_3 = 0$$

- Element deformations due to $D_3=1$

$$d_1 = 0, \quad d_2 = \cos\theta = 0.6, \quad d_3 = 1$$

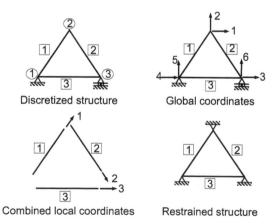

FIGURE 4.7f
Discretized structure and coordinates—Structure III.

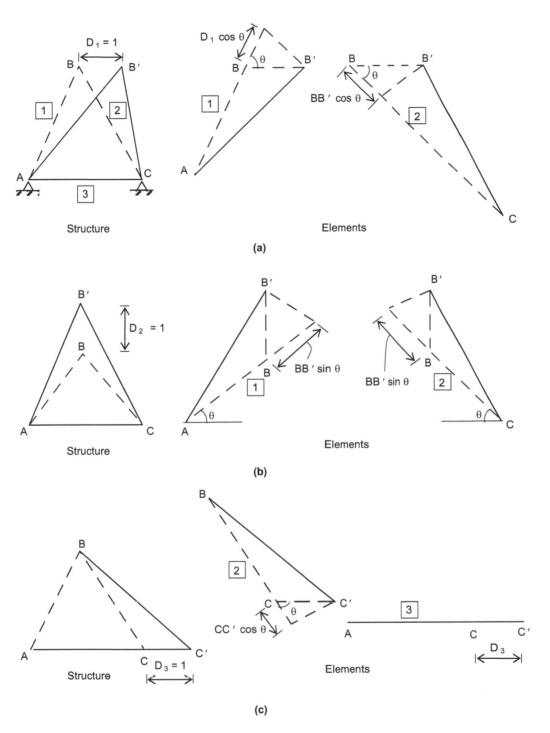

FIGURE 4.7g
Deformations of structure and elements: (a) $D_1 = 1$, (b) $D_2 = 1$, (c) $D_3 = 1$, (d) $D_4 = 1$, (e) $D_5 = 1$, and
(f) $D_6 = 1$.

(Continued)

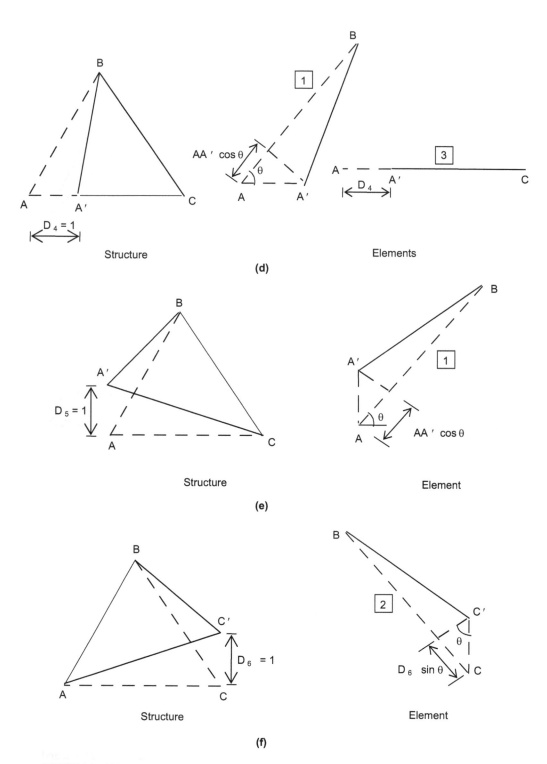

FIGURE 4.7g (CONTINUED)
Deformations of structure and elements: (a) $D_1 = 1$, (b) $D_2 = 1$, (c) $D_3 = 1$, (d) $D_4 = 1$, (e) $D_5 = 1$, and (f) $D_6 = 1$.

- Element deformations due to $D_4 = 1$
$$d_1 = -\cos\theta = -0.6, \quad d_2 = 0, \quad d_3 = -1$$
- Element deformations due to $D_5 = 1$
$$d_1 = -\sin\theta = -0.8, \quad d_2 = 0, \quad d_3 = 0$$
- Element deformations due to $D_6 = 1$
$$d_1 = 0, \quad d_2 = -\sin\theta = -0.8, \quad d_3 = 0$$

Displacement transformation matrix $[T_D]$

The global coordinates 1 to 3 are active coordinates and 4 to 6 are restrained coordinates.

- $[T_{DA}]$

$$[T_{DA}] = \begin{bmatrix} \overset{D_1=1}{0.6} & \overset{D_2=1}{0.8} & \overset{D_3=1}{0} \\ -0.6 & 0.8 & 0.6 \\ 0 & 0 & 1 \end{bmatrix}$$

- $[T_{DR}]$

$$[T_{DR}] = \begin{bmatrix} \overset{D_4=1}{-0.6} & \overset{D_5=1}{-0.8} & \overset{D_6=1}{0} \\ 0 & 0 & -0.8 \\ -1 & 0 & 0 \end{bmatrix}$$

- $[T_D]$

$$[T_D] = \begin{bmatrix} [T_{DA}] & \vdots & [T_{DR}] \end{bmatrix}$$

4.4 Equilibrium Matrix and the Principle of Contragradience

4.4.1 Equilibrium Matrix

The nodal forces $\{P\}$ can be written in terms of element forces $\{p\}$ using the equations of equilibrium as

$$\{P\} = [e]\{p\} \tag{4.8a}$$

where $[e]$ is the *equilibrium matrix*.

$[e]$ is a transformation matrix that transforms forces from local coordinates to global coordinates. The coefficients in $[e]$ are obtained using the FBD of elements and nodes. To indicate the influence of active and restrained coordinates, $[e]$ is partitioned into two sub-matrices $[e_A]$ and $[e_R]$. Equation 4.8a can be written as

$$\{P_A\} = [e_A]\{p\} \tag{4.8b}$$

$$\{P_R\} = [e_R]\{p\} \tag{4.8c}$$

4.4.2 Principle of Contragradience

The vectors $\{p\}$ and $\{d\}$ represent forces and displacements of the elements in local coordinates. Hence, the internal work done or the strain energy of the whole structure is $1/2\{p\}^T\{d\}$.

Similarly, $\{P\}$ and $\{D\}$ are the nodal forces and displacements of the structure (i.e., in global coordinates). Thus, the external work is $1/2\{P\}^T\{D\}$. Equating external work with the strain energy gives

$$\frac{1}{2}\{P\}^T\{D\} = \frac{1}{2}\{p\}^T\{d\}$$ (a)

From Eq. 4.6, $\{d\} = [T_D]\{D\}$. Hence,

$$\frac{1}{2}\{P\}^T\{D\} = \frac{1}{2}\{p\}^T[T_D]\{D\}$$

$$\Rightarrow \{P\}^T = \{p\}^T[T_D]$$ (b)

From Eq. 4.8,

$$\{P\} = [e]\{p\} \Rightarrow \{P\}^T = \{p\}^T[e]^T$$

Substituting in Eq. (b) gives

$$\{p\}^T[e]^T = \{p\}^T[T_D]$$

Hence,

$$[T_D] = [e]^T$$ (4.9a)

or

$$[e] = [T_D]^T$$ (4.9b)

The above relation between the two transformation matrices $[T_D]$ and $[e]$ is known as the *principle of contragradience*, which states that if four vectors are related by the equation $\{P\}^T\{D\} = \{p\}^T\{d\}$, then one transformation $\{P\} = [e]\{p\}$ will lead to another transformation $\{d\} = [e]^T\{D\}$.

Equation 4.9 can be used to develop the displacement transformation matrix using the equilibrium matrix.

Example 4.2

Establish the principle of contragradience for the system of springs shown in Figure 4.8a.

Solution

The discretized structure, restrained structure, and coordinates are shown in Figure 4.8b.

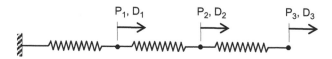

FIGURE 4.8a
System of springs.

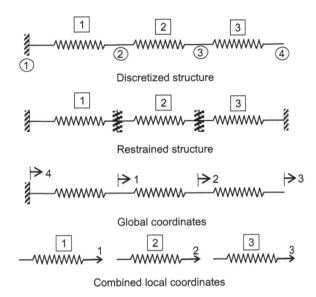

FIGURE 4.8b
Discretized structure and coordinates.

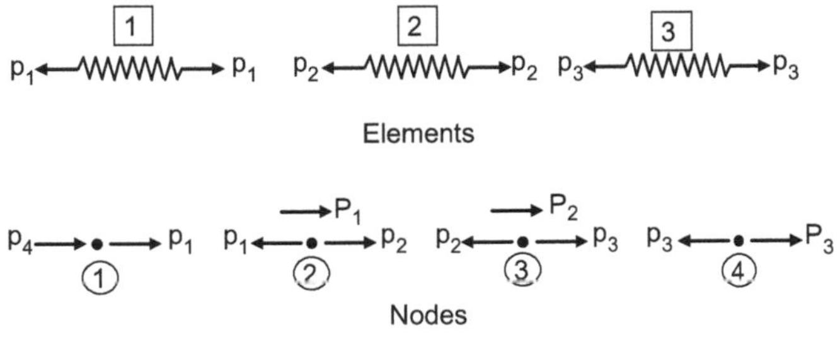

FIGURE 4.8c
FBD of elements and nodes.

- Equilibrium matrix [e]
  The FBD of elements and nodes are shown in Figure 4.8c.
  From the FBD of nodes,

$$P_1 = p_1 - p_2$$

$$P_2 = p_2 - p_3$$

$$P_3 = p_3$$

$$P_4 = -p_1$$

The above equations in the matrix form give

$$
\begin{Bmatrix} P_1 \\ P_2 \\ P_3 \\ P_4 \end{Bmatrix} = \begin{bmatrix} 1 & -1 & 0 \\ 0 & 1 & -1 \\ 0 & 0 & 1 \\ -1 & 0 & 0 \end{bmatrix} \begin{Bmatrix} p_1 \\ p_2 \\ p_3 \end{Bmatrix}
$$

Matrix Methods of Structural Analysis

or

$$\{P\} = [e]\{p\}$$

$$\therefore \ [e] = \begin{bmatrix} 1 & -1 & 0 \\ 0 & 1 & -1 \\ 0 & 0 & 1 \\ -1 & 0 & 0 \end{bmatrix}$$

- Displacement transformation matrix $[T_D]$

  The element deformations are obtained by applying unit displacements in the restrained structure.

i. Element deformations due to $D_1 = 1$

$$d_1 = 1, \quad d_2 = -1, \quad d_3 = 0$$

ii. Element deformations due to $D_2 = 1$

$$d_1 = 0, \quad d_2 = 1, \quad d_3 = -1$$

iii. Element deformations due to $D_3 = 1$

$$d_1 = 0, \quad d_2 = 0, \quad d_3 = 1$$

iv. Element deformations due to $D_4 = 1$

$$d_1 = -1, \quad d_2 = 0, \quad d_3 = 0$$

Thus,

$$[T_D] = \begin{bmatrix} \overset{D_1=1}{1} & \overset{D_2=1}{0} & \overset{D_3=1}{0} & \overset{D_4=1}{-1} \\ -1 & 1 & 0 & 0 \\ 0 & -1 & 1 & 0 \end{bmatrix}$$

It can be observed that $[T_D] = [e]^T$, which is the principle of contragradience.

Example 4.3

Develop the displacement transformation matrix for the structure shown in Figure 4.7a using the principle of contragradience.

Solution

Structure I

The forces acting on coordinates, and the FBD of elements and nodes, are shown in Figure 4.9a.

From the FBD of nodes,

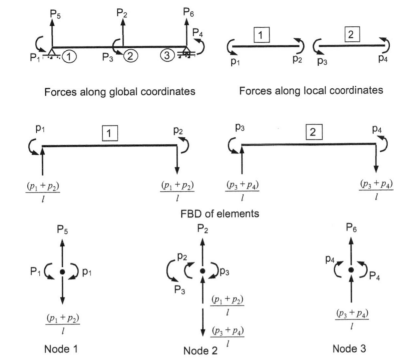

FIGURE 4.9a
Structure I.

$$P_1 = p_1$$

$$P_2 = -\frac{p_1}{l} - \frac{p_2}{l} + \frac{p_3}{l} + \frac{p_4}{l}$$

$$P_3 = p_2 + p_3$$

$$P_4 = p_4$$

$$P_5 = \frac{p_1}{l} + \frac{p_2}{l}$$

$$P_6 = -\frac{p_3}{l} - \frac{p_4}{l}$$

The equations in matrix form is

$$\left\{ \begin{array}{c} P_1 \\ P_2 \\ P_3 \\ P_4 \end{array} \right\} = \left[\begin{array}{cccc} 1 & 0 & 0 & 0 \\ -1/l & -1/l & 1/l & 1/l \\ 0 & 1 & 1 & 0 \\ 0 & 0 & 0 & 1 \end{array} \right] \left\{ \begin{array}{c} p_1 \\ p_2 \\ p_3 \\ p_4 \end{array} \right\}$$

or

$$\{P_A\} = [e_A]\{p\}$$

where $[e_A]$ is the equilibrium matrix corresponding to active coordinates and

$$\begin{Bmatrix} P_5 \\ P_6 \end{Bmatrix} = \begin{bmatrix} 1/l & 1/l & 0 & 0 \\ 0 & 0 & -1/l & -1/l \end{bmatrix} \begin{Bmatrix} p_1 \\ p_2 \\ p_3 \\ p_4 \end{Bmatrix}$$

or

$$\{P_R\} = [e_R]\{p\}$$

where $[e_R]$ is the equilibrium matrix corresponding to restrained coordinates.

- Displacement transformation matrix $[T_D]$

$$[T_{DA}] = [e_A]^T = \begin{bmatrix} 1 & -1/l & 0 & 0 \\ 0 & -1/l & 1 & 0 \\ 0 & 1/l & 1 & 0 \\ 0 & 1/l & 0 & 1 \end{bmatrix}$$

$$[T_{DR}] = [e_R]^T = \begin{bmatrix} 1/l & 0 \\ 1/l & 0 \\ 0 & -1/l \\ 0 & -1/l \end{bmatrix}$$

$$[T_D] = \begin{bmatrix} [T_{DA}] & \vdots & [T_{DR}] \end{bmatrix}$$

Structure II

The forces along the global and local coordinates, and the FBD of the elements and nodes, are shown in Figure 4.9b. In the figure, N_1, N_2, and N_3 are the axial forces in the elements.

From the FBD of nodes,

$$N_2 = \frac{(p_1 + p_2)}{l}$$

$$P_1 = N_2 + \frac{(p_5 + p_6)}{l} = \frac{p_1}{l} + \frac{p_2}{l} + \frac{p_5}{l} + \frac{p_6}{l} \tag{a}$$

$$P_2 = p_2 + p_3 \tag{b}$$

$$P_3 = p_4 + p_5 \tag{c}$$

$$P_4 = p_6 \tag{d}$$

$$P_5 = -\frac{p_1}{l} - \frac{p_2}{l} \tag{e}$$

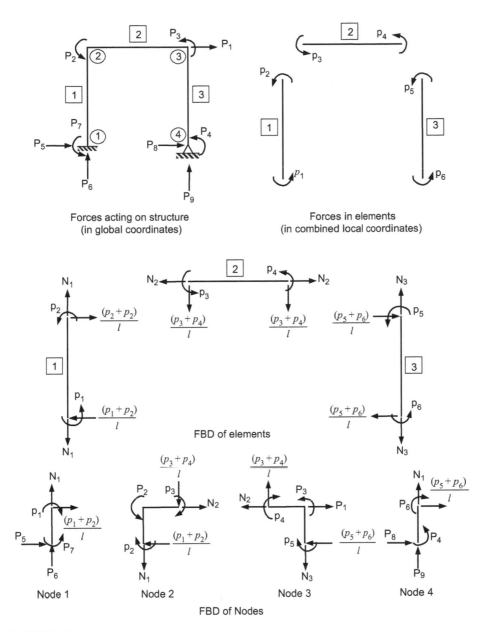

Forces acting on structure
(in global coordinates)

Forces in elements
(in combined local coordinates)

FBD of elements

Node 1   Node 2   Node 3   Node 4

FBD of Nodes

FIGURE 4.9b
Structure II.

$$N_1 = \frac{-(p_3 + p_4)}{l}$$

$$P_6 = -N_1 = \frac{p_3}{l} + \frac{p_4}{l} \tag{f}$$

$$P_7 = p_1 \tag{g}$$

$$P_8 = -\frac{p_5}{l} - \frac{p_6}{l} \tag{h}$$

$$N_3 = \frac{(p_3 + p_4)}{l}$$

$$P_9 = -N_3 = -\frac{p_3}{l} - \frac{p_4}{l} \qquad\qquad \text{(i)}$$

Equations (a) to (i) in the matrix form give

$$\{P_A\} = [e_A]\{p\}$$

$$\Rightarrow \begin{Bmatrix} P_1 \\ P_2 \\ P_3 \\ P_4 \end{Bmatrix} = \begin{bmatrix} 1/l & 1/l & 0 & 0 & 1/l & 1/l \\ 0 & 1 & 1 & 0 & 0 & 0 \\ 0 & 0 & 0 & 1 & 1 & 0 \\ 0 & 0 & 0 & 0 & 0 & 1 \end{bmatrix} \begin{Bmatrix} p_1 \\ p_2 \\ p_3 \\ p_4 \\ p_5 \\ p_6 \end{Bmatrix}$$

$$\{P_R\} = [e_R]\{p\}$$

$$\Rightarrow \begin{Bmatrix} P_5 \\ P_6 \\ P_7 \\ P_8 \\ P_9 \end{Bmatrix} = \begin{bmatrix} -1/l & -1/l & 0 & 0 & 0 & 0 \\ 0 & 0 & 1/l & 1/l & 0 & 0 \\ 1 & 0 & 0 & 0 & 0 & 0 \\ 0 & 0 & 0 & 0 & -1/l & -1/l \\ 0 & 0 & -1/l & -1/l & 0 & 0 \end{bmatrix} \begin{Bmatrix} p_1 \\ p_2 \\ p_3 \\ p_4 \\ p_5 \\ p_6 \end{Bmatrix}$$

The displacement transformation matrix of the structure is obtained by taking transpose of equilibrium matrix.

$$[T_{DA}] = [e_A]^T = \begin{bmatrix} 1/l & 0 & 0 & 0 \\ 1/l & 1 & 0 & 0 \\ 0 & 1 & 0 & 0 \\ 0 & 0 & 1 & 0 \\ 1/l & 0 & 1 & 0 \\ 1/l & 0 & 0 & 1 \end{bmatrix}$$

$$[T_{DR}] = [e_R]^T = \begin{bmatrix} -1/l & 0 & 1 & 0 & 0 \\ -1/l & 0 & 0 & 0 & 0 \\ 0 & 1/l & 0 & 0 & -1/l \\ 0 & 1/l & 0 & 0 & -1/l \\ 0 & 0 & 0 & -1/l & 0 \\ 0 & 0 & 0 & -1/l & 0 \end{bmatrix}$$

$$[T_D] = \begin{bmatrix} [T_{DA}] & \vdots & [T_{DR}] \end{bmatrix}$$

Structure III

The forces in different coordinates and the FBD of elements and nodes are shown in Figure 4.9c.

Using the FBD of nodes,

$$P_1 = p_1 \cos\theta - p_2 \cos\theta = 0.6p_1 - 0.6p_2 \tag{j}$$

$$P_2 = p_1 \sin\theta + p_2 \sin\theta = 0.8p_1 + 0.8p_2 \tag{k}$$

$$P_3 = p_2 \cos\theta + p_3 = 0.6p_2 + p_3 \tag{l}$$

$$P_4 = -p_1 \cos\theta - p_3 = -0.6p_1 - p_3 \tag{m}$$

$$P_5 = -p_1 \sin\theta = -0.8p_1 \tag{n}$$

$$P_6 = -p_2 \sin\theta = -0.8p_2 \tag{o}$$

Equations (j) to (o) in matrix form give

$$\{P_A\} = [e_A]\{p\}$$

$$\Rightarrow \begin{Bmatrix} P_1 \\ P_2 \\ P_3 \end{Bmatrix} = \begin{bmatrix} 0.6 & -0.6 & 0 \\ 0.8 & 0.8 & 0 \\ 0 & 0.6 & 1 \end{bmatrix} \begin{Bmatrix} p_1 \\ p_2 \\ p_3 \end{Bmatrix}$$

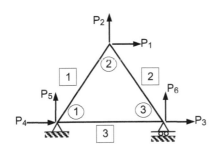

Forces acting on structure
(in global coordinates)

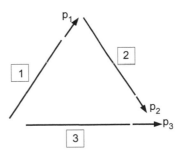

Forces in elements
(in combined local coordinates)

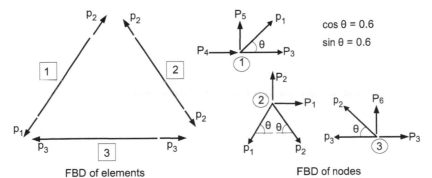

FBD of elements

FBD of nodes

$\cos\theta = 0.6$

$\sin\theta = 0.6$

FIGURE 4.9c
Structure III.

$$\{P_R\} = [e_R]\{p\}$$

$$\Rightarrow \begin{Bmatrix} P_4 \\ P_5 \\ P_6 \end{Bmatrix} = \begin{bmatrix} -0.6 & 0 & -1 \\ -0.8 & 0 & 0 \\ 0 & -0.8 & 0 \end{bmatrix} \begin{Bmatrix} p_1 \\ p_2 \\ p_3 \end{Bmatrix}$$

Hence,

$$[T_{DA}] = [e_A]^T = \begin{bmatrix} 0.6 & 0.8 & 0 \\ -0.6 & 0.8 & 0.6 \\ 0 & 0 & 1 \end{bmatrix}$$

$$[T_{DR}] = [e_R]^T = \begin{bmatrix} -0.6 & -0.8 & 0 \\ 0 & 0 & -0.8 \\ -1 & 0 & 0 \end{bmatrix}$$

$$[T_D] = \begin{bmatrix} [T_{DA}] & \vdots & [T_{DR}] \end{bmatrix}$$

Comments

From this example, it is observed that developing $[e]$ for a truss is easier when compared to developing $[T_D]$ using the deformed shape of truss. Hence for trusses, Eq. 4.9 will be used for getting $[T_D]$ from $[e]$. However, for beams and frames, it will be easier to develop $[T_D]$ using the deformed shape of the structure rather than getting $[T_D]$ from $[e]$.

4.5 Force–Displacement Relations

The force–displacement relations for the element and the structure used in the stiffness method are given below.

- For an element

  The force–displacement relation for the element i is

  $$\{p^i\} = [k^i]\{d^i\} \tag{4.10}$$

  where $\{p^i\}$, $\{d^i\}$ and $[k^i]$ are the force vector, displacement vector, and element stiffness matrix of the ith element of the structure in local coordinates.

  To reduce computational time, only basic element stiffness matrix will be used in this chapter (see Eq. 4.5).

- For the unassembled structure

  The force–displacement relations of the elements (say n number of elements) of the structure are combined to get the force–displacement relation of the unassembled structure. Thus,

$$\{p^1\}=[k^1]=\{d^1\}$$
$$\{p^2\}=[k^2]=\{d^2\}$$

....

$$\{p^n\}=[k^n]=\{d^n\}$$

The above equations in matrix form give

$$
\begin{Bmatrix} \{p^1\} \\ \{p^2\} \\ \cdots \\ \{p^n\} \end{Bmatrix} =
\begin{bmatrix} [k^1] & 0 & 0 & \cdots & 0 \\ 0 & [k^2] & 0 & \cdots & 0 \\ \cdots & \cdots & \cdots & \cdots & \cdots \\ 0 & 0 & 0 & \cdots & [k^n] \end{bmatrix}
\begin{Bmatrix} \{d^1\} \\ \{d^2\} \\ \cdots \\ \{d^n\} \end{Bmatrix}
$$

$$\Rightarrow \quad \{p\}=[k]\{d\} \tag{4.11}$$

where $\{p\}$, $\{d\}$, and $[k]$ are the combined element force and displacement vectors and

$$
[k] =
\begin{bmatrix} [k^1] & 0 & 0 & \cdots & 0 \\ 0 & [k^2] & 0 & \cdots & 0 \\ \cdots & \cdots & \cdots & \cdots & \cdots \\ 0 & 0 & 0 & \cdots & [k^n] \end{bmatrix} \tag{4.12}
$$

is a diagonal matrix with the element stiffness matrix as its constituents. $[k]$ is known as the *unassembled stiffness matrix*.

- For the structure
  The force–displacement relation for the structure is

$$\{P\}=[K]\{D\} \tag{4.13}$$

where $\{P\}$ and $\{D\}$ are the nodal force and displacement vectors of the structure. $[K]$ is the *stiffness matrix* of the structure or the *structure stiffness matrix*.

To indicate the effects of active and restrained coordinates, Eq. 4.13 can be written as

$$\{P\}=[K]\{D\}$$

or

$$
\begin{Bmatrix} \{P_A\} \\ \hline \{P_R\} \end{Bmatrix} =
\begin{bmatrix} [K_{AA}] & [K_{AR}] \\ \hline [K_{RA}] & [K_{RR}] \end{bmatrix}
\begin{Bmatrix} \{D_A\} \\ \hline \{D_R\} \end{Bmatrix} \tag{4.14}
$$

Thus,

$$\{P_A\} = [K_{AA}]\{D_A\} + [K_{RA}]\{D_R\} \tag{4.15a}$$

$$\{P_R\} = [K_{RA}]\{D_A\} + [K_{RR}]\{D_R\} \tag{4.15b}$$

4.6 Structure Stiffness Matrix [K]

The law of conservation of energy is used to derive the expression for the structure stiffness matrix [K]. The work done by the external load W is

$$W = \frac{1}{2}\{D\}^T\{P\} = \frac{1}{2}\{D\}^T[K]\{D\} \quad (\because \{P\} = [K]\{D\})$$

The work done by the internal forces or the strain energy of the structure U is

$$U = \frac{1}{2}\{d\}^T\{p\}$$

But $\{d\} = [T_D]\,\{D\}$

$$\Rightarrow \{d\}^T = \{D\}^T[T_D]^T$$

Also $\{p\} = [k]\,\{d\} = [k]\,[T_D]\,\{D\}$

Therefore, $U = \frac{1}{2}\{d\}^T\{p\} = \frac{1}{2}\{D\}^T[T_D]^T[k][T_D]\{D\}$

Equating W and U gives

$$W = U$$

$$\Rightarrow \frac{1}{2}\{D\}^T[K]\{D\} = \frac{1}{2}\{D\}^T[T_D]^T[k][T_D]\{D\}$$

Thus,

$$[K] = [T_D]^T[k][T_D] \tag{4.16}$$

But

$$[T_D] = \left[\ \ [T_{DA}] \ \vdots \ [T_{DR}] \ \ \right]$$

$$\Rightarrow [T_D]^T = \left[\begin{array}{c} [T_{DA}]^T \\ \hline [T_{DR}]^T \end{array}\right]$$

Therefore, $[K] = \begin{bmatrix} [T_{DA}]^T \\ \hline [T_{DR}]^T \end{bmatrix} [k] \begin{bmatrix} T_{DA} & \vdots & T_{DR} \end{bmatrix}$

$$= \begin{bmatrix} [T_{DA}]^T [k][T_{DA}] & \vdots & [T_{DA}]^T [k][T_{DR}] \\ \hline [T_{DR}]^T [k][T_{DA}] & \vdots & [T_{DR}]^T [k][T_{DR}] \end{bmatrix} \qquad (4.17)$$

But $[K]$ is partitioned as (Eq. 4.14),

$$[K] = \begin{bmatrix} [K_{AA}] & \vdots & [K_{AR}] \\ \hline [K_{RA}] & \vdots & [K_{RR}] \end{bmatrix}$$

Hence,

$$[K_{AA}] = [T_{DA}]^T [k][T_{DA}] \qquad (4.18a)$$

$$[K_{AR}] = [T_{DA}]^T [k][T_{DR}] \qquad (4.18b)$$

$$[K_{RA}] = [T_{DR}]^T [k][T_{DA}] \qquad (4.18c)$$

$$[K_{RR}] = [T_{DR}]^T [k][T_{DR}] \qquad (4.18d)$$

- Relation between $[K_{AR}]$ and $[K_{RA}]$
  From Eq. 4.18b,

$$[K_{AR}] = [T_{DA}]^T [k][T_{DR}]$$

Therefore, $[K_{AR}]^T = [T_{DR}]^T [k]^T [T_{DA}]$

But $[k]^T = [k]$ (since $[k]$ is a diagonal matrix)

$$\Rightarrow [K_{AR}]^T = [T_{DR}]^T [k][T_{DA}] = [K_{RA}]$$

Thus,

$$[K_{RA}] = [K_{AR}]^T \qquad (4.19)$$

4.7 Equilibrium Equation

The equilibrium equation used in the stiffness method of analysis is

$$\{P\} = [K]\{D\}$$

From Eq. (4.14),

$$\left\{ \begin{array}{c} \{P_A\} \\ \hline \{P_R\} \end{array} \right\} = \left[\begin{array}{cc} [K_{AA}] & [K_{AR}] \\ \hline [K_{RA}] & [K_{RR}] \end{array} \right] \left\{ \begin{array}{c} \{D_A\} \\ \hline \{D_R\} \end{array} \right\}$$

In the above equation, $\{P_A\}$ and $\{D_R\}$ are known vectors. $\{D_A\}$ and $\{P_R\}$ are unknowns. $\{D_A\}$ is primary unknown quantity in the stiffness method and is determined using Eq. 4.15a.

$$\{P_A\} = [K_{AA}]\{D_A\} + [K_{AR}]\{D_R\}$$

$$\Rightarrow \quad \{D_A\} = [K_{AA}]^{-1}\left(\{P_A\} - [K_{AR}]\{D_R\}\right) \qquad (4.20)$$

After finding $\{D_A\}$, $\{P_R\}$ is calculated using Eq. 4.15b.

4.8 Transformations Used in Stiffness Method

Figure 4.10 shows the various transformations used in the stiffness method. From Figure 4.10,

$$\{d\} = [T_D]\{D\}$$

$$\{p\} = [k]\{d\} = [k][T_D]\{D\}$$

$$\{P\} = [e]\{p\} = [T_D]^T\{p\} = [T_D]^T[k][T_D]\{D\}$$

$$\text{But } \{P\} = [K]\{D\} = [T_D]^T[k][T_D]\{D\}$$

Thus[*],

$$[K] = [T_D]^T[k][T_D]$$

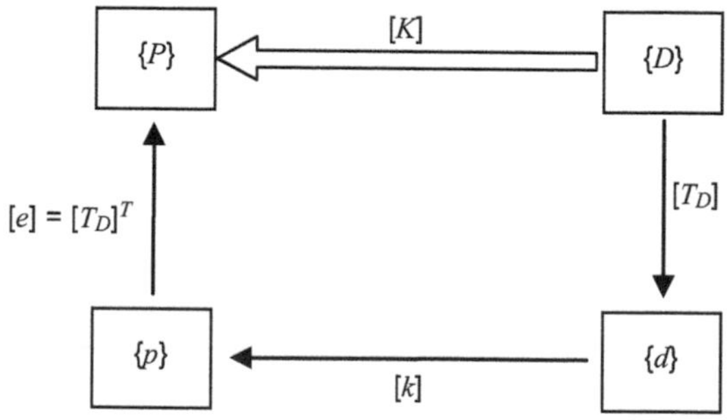

FIGURE 4.10
Transformation matrices in the stiffness method.

---

[*] This is an alternate proof for finding the expression for [K].

4.9 Procedure for Analysis of Structures Using Stiffness Method

The steps required for the analysis of structures using the stiffness method are listed below:

  i. Develop the discretized structure and number the nodes and elements.
 ii. Identify the active and restrained coordinates (global coordinates).
iii. Define the load vector $\{P_A\}$, reaction vector $\{P_R\}$, active displacement vector $\{D_A\}$, and restrained displacement vector $\{D_R\}$.
 iv. Identify the local coordinates and combined local coordinates. Define the combined element force and displacement vectors ($\{p\}$ and $\{d\}$).
  v. Determine the displacement transformation matrix

$$[T_D]=\left[\ \ [T_{DA}]\ \vdots\ [T_{DR}]\ \ \right]$$

 vi. Determine the element stiffness matrix $[k^i]$ for all the elements and assemble them in a diagonal matrix to get the unassembled stiffness matrix $[k]$
vii. Compute the structure stiffness matrix $[K]$ using Eq. 4.17.
viii. Find the active displacement vector $\{D_A\}$ using Eq. 4.20.
 ix. After finding $\{D_A\}$, find the reaction vector using Eq. 4.15b.
  x. The element forces are determined using Eqs. 4.11 and 4.7.

$$\{p\}=[k]\{d\}=[k]([T_{DA}]\{D_A\}+[T_{DR}]\{D_R\})$$

$$\text{Thus,}\quad \{p\}=[k][T_{DA}]\{D_A\}+[k][T_{DR}]\{D_R\} \tag{4.21}$$

4.9.1 Analysis of Structures with Element Loads

In these types of structures, the combined force vectors along the active and restrained coordinates $\{P_{CA}\}$ and $\{P_{CR}\}$ are to be used instead of $\{P_A\}$ and $\{P_R\}$. Using the principle of superposition, the force in the elements $\{p\}$ are obtained as

$$\{p\}=\{p_F\}+\{p_C\} \tag{4.22}$$

where $\{p_F\}$ are the element forces due to fixed end actions and $\{p_C\}$ are the element forces due to combined nodal loads calculated using Eq. 4.21.

4.9.2 Alternate Simplified Method of Analysis

In structures where there is no settlement of supports or when the effects of support settlements are converted into equivalent fixed end actions, then for such cases, $\{D_R\} = \{0\}$.
Hence, from Eqs. 4.20 and 4.21,

$$\{D_A\}=[K_{AA}]^{-1}\{P_A\} \tag{4.23}$$

$$\{p\}=[k][T_{DA}]\{D_A\} \tag{4.24}$$

After finding the element forces, the support reactions are calculated using the FBD of the elements and nodes.

The advantage of using this approach in analysis is that only active coordinates need to be considered. Hence, only $[T_{DA}]$ and $[K_{AA}]$ have to be calculated. Thus, less time is required for analysis (see Example 4.4).

Example 4.4

Analyze the structure made of axial members shown in Figure 4.11a using the stiffness method.

Solution

Step 1: Develop the discretized structure and identify the coordinates
  The discretized structure, coordinates, and restrained structure are shown in Figure 4.11b.
Step 2: Identify the force and displacement vectors of the structure.

- Load vector $\{P_A\}$

$$\{P_A\} = \left\{ \begin{array}{c} P_1 \\ P_2 \\ P_3 \end{array} \right\} = \left\{ \begin{array}{c} 60 \\ 10 \\ 30 \end{array} \right\} \text{kN}$$

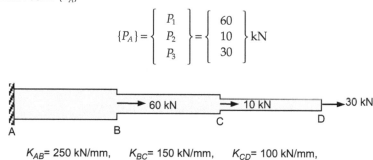

K_{AB}= 250 kN/mm, K_{BC}= 150 kN/mm, K_{CD}= 100 kN/mm,

FIGURE 4.11a
Example 4.4.

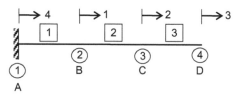

Discretized structure and global coordinates

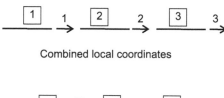

Combined local coordinates

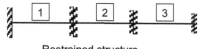

Restrained structure

FIGURE 4.11b
Elements, nodes, coordinates, and restrained structure.

- Reaction vector $\{P_R\}$

$$\{P_R\} = \{P_4\}$$

- Active displacement vector $\{D_A\}$

$$\{D_A\} = \left\{ \begin{array}{c} D_1 \\ D_2 \\ D_3 \end{array} \right\}$$

- Restrained displacement vector $\{D_R\}$

$$\{D_R\} = \{D_4\} = \{0\}$$

Step 3: Develop displacement transformation matrix $[T_D]$

Using the restrained structure, $[T_D]$ is developed by applying unit displacement along the global coordinates one at a time. The structure is similar to the system of springs considered in Example 4.2. Hence,

$$[T_{DA}] = \begin{bmatrix} 1 & 0 & 0 \\ -1 & 1 & 0 \\ 0 & -1 & 1 \end{bmatrix} \quad [T_{DR}] = \left\{ \begin{array}{c} -1 \\ 0 \\ 0 \end{array} \right\}$$

Step 4: Develop the structure stiffness matrix

- Element stiffness matrix $[k^i]$

$$k^1 = K_{AB} = 250 \text{ kN/mm}, \quad k^2 = K_{BC} = 150 \text{ kN/mm}, \quad \text{and } k^3 = K_{CD} = 100 \text{ kN/mm}$$

- Unassembled stiffness matrix $[k]$

$$[k] = \begin{bmatrix} k^1 & 0 & 0 \\ 0 & k^2 & 0 \\ 0 & 0 & k^3 \end{bmatrix} = \begin{bmatrix} 250 & 0 & 0 \\ 0 & 150 & 0 \\ 0 & 0 & 100 \end{bmatrix} \text{kN/mm}$$

- $[K_{AA}], [K_{AR}], [K_{RA}], [K_{RR}]$

$$[K_{AA}] = [T_{DA}]^T [k][T_{DA}] = \begin{bmatrix} 400 & -150 & 0 \\ -150 & 250 & -100 \\ 0 & -100 & 100 \end{bmatrix}$$

$$[K_{AR}] = [T_{DA}]^T [k][T_{DR}] = \left\{ \begin{array}{c} -250 \\ 0 \\ 0 \end{array} \right\}$$

$$[K_{RA}] = [K_{AR}]^T = \begin{bmatrix} -250 & 0 & 0 \end{bmatrix}$$

$$[K_{RR}] = [T_{DR}]^T [k][T_{DR}] = [250]$$

Step 5: Find $\{D_A\}$

- Element flexibility matrix

$$\{D_A\} = [K_{AA}]^{-1}\left(\{P_A\} - [K_{AR}]\{D_R\}\right) = \left\{ \begin{array}{c} 0.4 \\ 0.667 \\ 0.9667 \end{array} \right\} \text{mm}$$

Step 6: Find $\{P_R\}$

$$\{P_R\} = [K_{RA}]\{D_A\} + [K_{RR}]\{D_R\} = \{-100\}$$

Step 7: Find $\{p\}$

$$\{p\} = [k][T_{DA}]\{D_A\} = \left\{ \begin{array}{c} 100 \\ 40 \\ 30 \end{array} \right\} \text{kN}$$

- Solution using the simplified method—by considering only active coordinates

For this method, $\{D_R\}$ should be a null vector. After finding $[T_{DA}]$, $[k]$, and $[K_{AA}]$, the nodal displacements $\{D_A\}$ are determined using the following equation:

$$\{D_A\} = [K_{AA}]^{-1}\{P_A\} = \left\{ \begin{array}{c} 0.4 \\ 0.667 \\ 0.9667 \end{array} \right\} \text{mm}$$

After finding $\{D_A\}$, the element force vector $\{p\}$ is calculated as

$$\{p\} = [k][T_{DA}]\{D_A\} = \left\{ \begin{array}{c} 100 \\ 40 \\ 30 \end{array} \right\} \text{kN}$$

The FBD of the elements and node 1 are shown in Figure 4.11c.

The support reactions are determined using the FBD of elements and nodes. From the FBD of node 1,

$$P_4 = -100 \text{ kN}$$

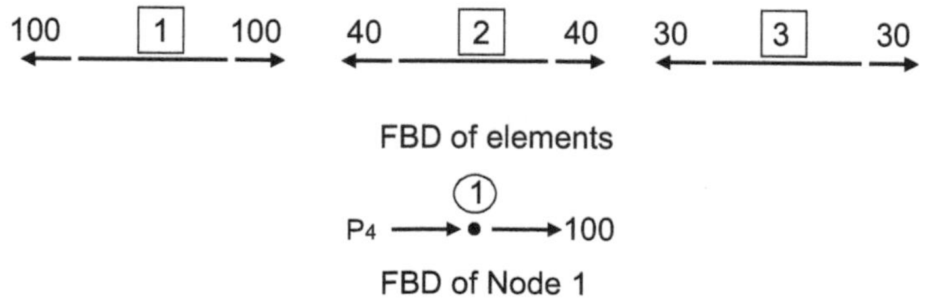

FBD of elements

FBD of Node 1

FIGURE 4.11c
FBD of elements and nodes.

The advantage of using the simplified method is that only active coordinates are to be considered. The restrained coordinates are not necessary for analysis, i.e., $[K_{AR}]$ and $[K_{RR}]$ need not be calculated. Hence, the analysis will take less time. The remaining examples in this chapter will be analyzed using this simplified method.

Example 4.5

Find the deflection at the free end of the beam shown in Figure 4.12a.

Solution

Step 1: Develop the discretized structure and identify the coordinates. The discretized structure, coordinates, and restrained structure are shown in Figure 4.12b.
Step 2: Identify the load vector.

$$\{P_A\} = \left\{ \begin{array}{c} P_1 \\ P_2 \end{array} \right\} = \left\{ \begin{array}{c} -P \\ 0 \end{array} \right\}$$

Step 3: Develop the displacement transformation matrix $[T_{DA}]$
   The deformed shape of the restrained structure due to the application of unit displacements at the global coordinates is shown in Figure 4.12c.
   From Figure 4.12c,

$$[T_{DA}] = \begin{array}{c} D_1=1 \quad D_2=1 \\ \left[\begin{array}{cc} -1/l & 0 \\ -1/l & 1 \end{array} \right] \end{array}$$

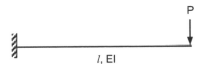

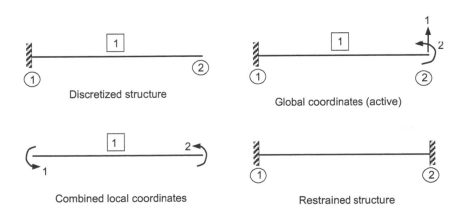

FIGURE 4.12a
Example 4.5.

Discretized structure

Global coordinates (active)

Combined local coordinates

Restrained structure

FIGURE 4.12b
Elements, nodes, coordinates, and restrained structure.

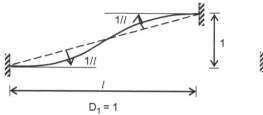
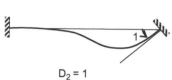

FIGURE 4.12c
Deformed shape of restrained structure.

Step 4: Develop the structure stiffness matrix $[K_{AA}]$

* Element stiffness matrix $[k^i]$

$$\left[k^1 \right] = \frac{EI}{l} \begin{bmatrix} 4 & 2 \\ 2 & 4 \end{bmatrix}$$

* Combined element stiffness matrix $[k]$

$$[k] = [k^1] = \frac{EI}{l} \begin{bmatrix} 4 & 2 \\ 2 & 4 \end{bmatrix}$$

* Structure stiffness matrix $[K_{AA}]$

$$[K_{AA}] = [T_{DA}]^T [k][T_{DA}] = \frac{EI}{l} \begin{bmatrix} 12/l^2 & -6/l \\ -6/l & 4 \end{bmatrix}$$

Step 5: Find the nodal displacements $\{D_A\}$

$$\{D_A\} = \left\{ \begin{array}{c} D_1 \\ D_2 \end{array} \right\}$$

$$\{D_A\} = [K_{AA}]^{-1}\{P_A\} = \left\{ \begin{array}{c} D_1 \\ D_2 \end{array} \right\} = \left\{ \begin{array}{c} -Pl^3/3EI \\ -Pl^2/2EI \end{array} \right\}$$

The deflection at the tip of the beam is $Pl^3/3EI(\downarrow)$

Example 4.6

Find the vertical deflection at joint B and the forces developed in the beam shown in Figure 4.13a.

Solution

Step 1: Develop the discretized structure

The discretized structure, coordinates, and restrained structure are shown in Figure 4.13b.

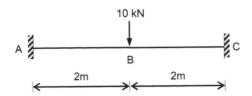

FIGURE 4.13a
Example 4.6.

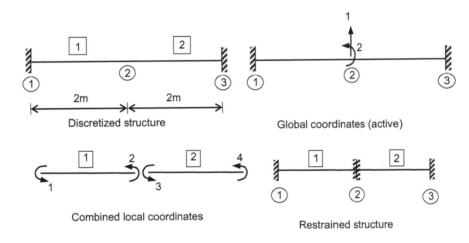

FIGURE 4.13b
Elements, nodes, coordinates, and restrained structure.

Step 2: Identify the load vector

$$\{P_A\} = \left\{ \begin{array}{c} P_1 \\ P_2 \end{array} \right\} = \left\{ \begin{array}{c} -10\text{kN} \\ 0 \end{array} \right\}$$

Step 3: Develop $[T_{DA}]$

The deformed shape of the restrained structure due to unit displacements at the global coordinates are shown in Figure 4.13c.

From Figure 4.13c,

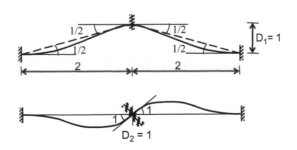

FIGURE 4.13c
Deformed shape of restrained structure.

$$[T_{DA}] = \begin{array}{cc} \scriptstyle D_1=1 & \scriptstyle D_2=1 \\ \left[\begin{array}{c:c} -1/2 & 0 \\ -1/2 & 1 \\ 1/2 & 1 \\ 1/2 & 0 \end{array} \right] \end{array}$$

Step 4: Develop $[K_{AA}]$

- Element stiffness matrix

$$[k^1] = [k^2] = \frac{EI}{l} \begin{bmatrix} 4 & 2 \\ 2 & 4 \end{bmatrix} = \frac{EI}{2} \begin{bmatrix} 4 & 2 \\ 2 & 4 \end{bmatrix}$$

- Combined element stiffness matrix $[k]$

$$[k] = \begin{bmatrix} [k^1] & [0] \\ [0] & [k^2] \end{bmatrix} = \frac{EI}{2} \begin{bmatrix} 4 & 2 & 0 & 0 \\ 2 & 4 & 0 & 0 \\ 0 & 0 & 4 & 2 \\ 0 & 0 & 2 & 4 \end{bmatrix}$$

- Structure stiffness matrix $[K_{AA}]$

$$[K_{AA}] = [T_{DA}]^T [k][T_{DA}] = \frac{EI}{2} \begin{bmatrix} 6 & 0 \\ 0 & 8 \end{bmatrix}$$

Step 5: Find $\{D_A\}$

$$\{D_A\} = [K_{AA}]^{-1}\{P_A\}$$

$$\{D_A\} = \begin{Bmatrix} D_1 \\ D_2 \end{Bmatrix} = \frac{2}{EI} \begin{Bmatrix} -1.667 \\ 0 \end{Bmatrix}$$

The vertical deflection at joint B is $3.334/EI$ ($\downarrow$)

Step 6: Find $\{p\}$

$$\{p\} = [k][T_{DA}]\{D_A\}$$

$$= \begin{Bmatrix} p_1 \\ p_2 \\ p_3 \\ p_4 \end{Bmatrix} = \begin{Bmatrix} 5 \\ 5 \\ -5 \\ -5 \end{Bmatrix}$$

The basic end actions of the elements and the FBD of the elements are shown in Figure 4.13d.

From the FBD of the elements,

$$V_A = 5, \quad V_{B,L} = -5, \quad V_{B,R} = -5, \quad V_C = 5 \text{ kN}$$

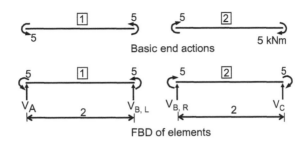

Basic end actions

FBD of elements

FIGURE 4.13d
Forces developed in elements.

Example 4.7

Analyze the structure shown in Figure 4.14a.

Solution

Step 1: Develop the discretized structure
The discretized structure, coordinates, and restrained structure are shown Figure 4.14b.
Step 2: Define the force vectors $\{P_{CA}\}$ and $\{p_F\}$
The element load is converted into nodal loads. The fixed end actions and equivalent nodal loads are shown in Figure 4.14c.

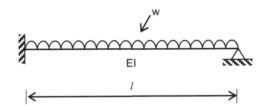

FIGURE 4.14a
Example 4.7.

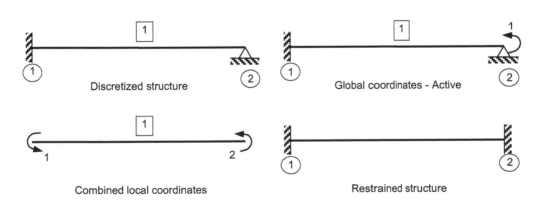

FIGURE 4.14b
Discretized structure, coordinates, and restrained structure.

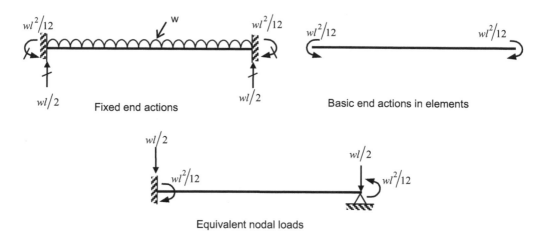

Fixed end actions   Basic end actions in elements

Equivalent nodal loads

FIGURE 4.14c
Fixed end actions, basic end actions, and equivalent nodal loads.

- Combined load vector $\{P_{CA}\}$

    Since there are no nodal loads acting on the structure, the equivalent nodal loads will be the combined nodal loads.

$$\{P_{CA}\} = \{P_{C1}\} = \left\{ \frac{wl^2}{12} \right\}$$

- Element forces due to fixed end actions $\{p_F\}$

    The basic end actions in elements caused by the fixed end actions are shown in Figure 4.14c.

$$\{p_F\} = \left\{ \begin{array}{c} wl^2/12 \\ -wl^2/12 \end{array} \right\}$$

Step 3: Develop $[T_{DA}]$

The deformed shape of the restrained structure is shown in Figure 4.14d. From the figure,

$$[T_{DA}] = \left[\begin{array}{c} 0 \\ 1 \end{array} \right]$$

Step 4: Develop $[K_{AA}]$

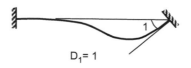

$D_1 = 1$

FIGURE 4.14d
Deformed shape of the restrained structure.

- Element stiffness matrix $[k^i]$

$$[k^1] = \frac{EI}{l} \begin{bmatrix} 4 & 2 \\ 2 & 4 \end{bmatrix}$$

- Unassembled stiffness matrix $[k]$

$$[k] = [k^1] = \frac{EI}{l} \begin{bmatrix} 4 & 2 \\ 2 & 4 \end{bmatrix}$$

- Structure stiffness matrix $[K_{AA}]$

$$[K_{AA}] = [T_{DA}]^T [k][T_{DA}] = \frac{4EI}{l}$$

Step 5: Find $\{D_A\}$

$$\{D_A\} = [K_{AA}]^{-1} \{P_A\}$$

$$\Rightarrow \quad D_1 = \frac{l}{4EI} \times \frac{wl^2}{12} = \frac{wl^3}{48EI}$$

Step 6: Find $\{p\}$

$$\{p\} = \{p_F\} + [k][T_{DA}]\{D_A\}$$

$$[k][T_{DA}]\{D_A\} = \left\{ \begin{array}{c} wl^2/24 \\ wl^2/12 \end{array} \right\}$$

$$\therefore \{p\} = \left\{ \begin{array}{c} wl^2/12 \\ -wl^2/12 \end{array} \right\} + \left\{ \begin{array}{c} wl^2/24 \\ wl^2/12 \end{array} \right\} = \left\{ \begin{array}{c} wl^2/8 \\ 0 \end{array} \right\}$$

Step 7: Find the forces in elements and support reactions. The FBD of structure, elements, and nodes are shown in Figure 4.14e.

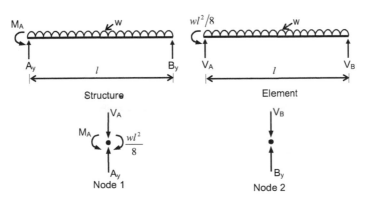

FIGURE 4.14e
FBD of structure, element, and nodes.

From the FBD of the element,

$$V_A = \frac{5}{8}wl, \ V_B = \frac{3}{8}wl$$

Once the element forces are determined, the SFD and BMD of the element and structure can be drawn.

The support reactions M_A, A_y, and B_y are determined using the FBD of the nodes. From the FBD of the nodes,

$$A_y = V_A = \frac{5}{8}wl, \quad M_A = \frac{wl^2}{8}, \quad B_y = V_B = \frac{3}{8}wl$$

Example 4.8

Analyze the structure shown in Figure 4.15a.

Solution

Step 1: Develop the discretized structure

Figure 4.15b shows the discretized structure, coordinates, and restrained structure.

Step 2: Define $\{P_{CA}\}$ and $\{p_F\}$

The fixed end actions and equivalent nodal loads are shown in Figure 4.15c.

$$\{P_{CA}\} = \begin{Bmatrix} P_{C1} \\ P_{C2} \end{Bmatrix} = \begin{Bmatrix} 1.5 \\ 2.5 \end{Bmatrix} \text{kNm}$$

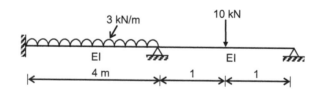

FIGURE 4.15a

Example 4.8.

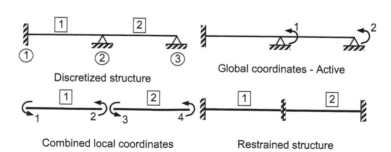

Discretized structure

Global coordinates - Active

Combined local coordinates

Restrained structure

FIGURE 4.15b

Discretized structure, coordinates, and restrained structure.

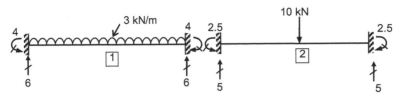

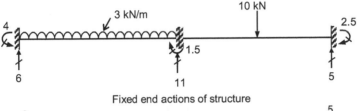

Fixed end actions of individual elements

Fixed end actions of structure

Equivalent nodal loads

FIGURE 4.15c
Fixed end actions and equivalent nodal loads.

$$\{p_F\} = \left\{ \begin{array}{c} 4 \\ -4 \\ 2.5 \\ -2.5 \end{array} \right\} \text{kNm}$$

Step 3: Develop $[T_{DA}]$
From Figure 4.15d,

$$[T_{DA}] = \begin{array}{cc} {\scriptstyle D_1=1 \quad D_2=1} \\ \begin{bmatrix} 0 & 0 \\ 1 & 0 \\ 1 & 0 \\ 0 & 1 \end{bmatrix} \end{array}$$

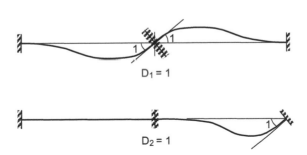

$D_1 = 1$

$D_2 = 1$

FIGURE 4.15d
Deformed shape of the restrained structure.

Step 4: Develop $[K_{AA}]$

- $[k^i]$

$$[k^1] = \frac{EI}{4}\begin{bmatrix} 4 & 2 \\ 2 & 4 \end{bmatrix} = EI\begin{bmatrix} 1 & 0.5 \\ 0.5 & 1 \end{bmatrix}$$

$$[k^2] = \frac{EI}{2}\begin{bmatrix} 4 & 2 \\ 2 & 4 \end{bmatrix} = EI\begin{bmatrix} 2 & 1 \\ 1 & 2 \end{bmatrix}$$

- $[k]$

$$[k] = \begin{bmatrix} [k^1] & [0] \\ [0] & [k^1] \end{bmatrix} = EI\begin{bmatrix} 1 & 0.5 & 0 & 0 \\ 0.5 & 1 & 0 & 0 \\ 0 & 0 & 2 & 1 \\ 0 & 0 & 1 & 2 \end{bmatrix}$$

- $[K_{AA}]$

$$[K_{AA}] = [T_{DA}]^T[k][T_{DA}] = EI\begin{bmatrix} 3 & 1 \\ 1 & 2 \end{bmatrix}$$

Step 5: Find $\{D_A\}$

$$\{D_A\} = [K_{AA}]^{-1}\{P_A\} = \frac{1}{EI}\begin{Bmatrix} 0.1 \\ 1.2 \end{Bmatrix}$$

Step 6: Find $\{p\}$

$$\{p\} = \{p_F\} + [k][T_{DA}]\{D_A\} = \begin{Bmatrix} 4.05 \\ -3.9 \\ 3.9 \\ 0 \end{Bmatrix} \text{kNm}$$

After finding $\{p\}$, the forces in the elements and the support reactions are determined using the FBD of elements and nodes (see Example 4.7).

Example 4.9

Analyze the beam shown in Figure 4.16a. The supports B and C sink by $100/EI$ and $50/EI$, respectively.

Solution

Step 1: Develop the discretized structure
   The element, nodes, coordinates, and restrained structure are shown Figure 4.16b.
Step 2: Find $\{P_{CA}\}$ and $\{p_F\}$
   The fixed end actions and combined nodal forces are shown in Figure 3.22r–t.

$$\{P_{CA}\} = \begin{Bmatrix} P_{C1} \\ P_{C2} \end{Bmatrix} = \begin{Bmatrix} -13.75 \\ 23.75 \end{Bmatrix} \text{kNm}$$

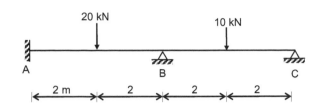

FIGURE 4.16a
Example 4.9.

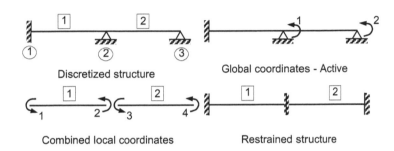

Discretized structure

Global coordinates - Active

Combined local coordinates

Restrained structure

FIGURE 4.16b
Discretized structure, coordinates, and restrained structure.

$$\{p_F\} = \begin{Bmatrix} 47.5 \\ 27.5 \\ -13.75 \\ -23.75 \end{Bmatrix} \text{kNm}$$

Step 3: Develop $[T_{DA}]$
  From Figure 4.16c,

$$[T_{DA}] = \begin{matrix} {\scriptstyle D_1=1 \quad D_2=1} \\ \begin{bmatrix} 0 & 0 \\ 1 & 0 \\ 1 & 0 \\ 0 & 1 \end{bmatrix} \end{matrix}$$

$D_1 = 1$

$D_2 = 1$

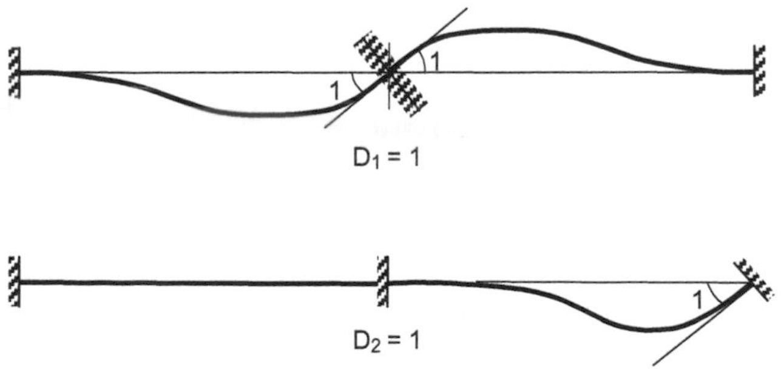

FIGURE 4.16c
Deformed shape of the restrained structure.

Step 4: Develop $[K_{AA}]$

- $[k^i]$

$$[k^1] = [k^2] = \frac{EI}{4}\begin{bmatrix} 4 & 2 \\ 2 & 4 \end{bmatrix} = EI\begin{bmatrix} 1 & 0.5 \\ 0.5 & 1 \end{bmatrix}$$

- $[k]$

$$[k] = \begin{bmatrix} [k^1] & [0] \\ [0] & [k^2] \end{bmatrix} = EI\begin{bmatrix} 1 & 0.5 & 0 & 0 \\ 0.5 & 1 & 0 & 0 \\ 0 & 0 & 2 & 1 \\ 0 & 0 & 1 & 2 \end{bmatrix}$$

- $[K_{AA}]$

$$[K_{AA}] = [T_{DA}]^T [k][T_{DA}] = EI\begin{bmatrix} 2 & 0.5 \\ 0.5 & 1 \end{bmatrix}$$

Step 5: Find $\{D_A\}$

$$\{D_A\} = [K_{AA}]^{-1}\{P_A\}$$

$$\Rightarrow \begin{Bmatrix} D_1 \\ D_2 \end{Bmatrix} = \frac{1}{EI}\begin{Bmatrix} -14.643 \\ 31.071 \end{Bmatrix}$$

Step 6: Find $\{p\}$

$$\{p\} = \{p_F\} + [k][T_{DA}]\{D_A\} = \begin{Bmatrix} 40.179 \\ 12.857 \\ -12.857 \\ 0 \end{Bmatrix}\text{kNm}$$

Example 4.10

Analyze the frame shown in Figure 4.17a.

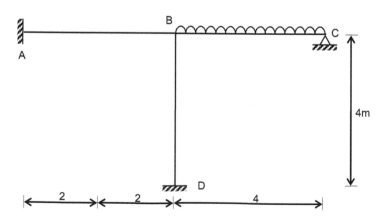

FIGURE 4.17a
Example 4.10.

Solution

Step 1: Develop the discretized structure

Figure 4.17b shows the discretized structure, coordinates, and restrained structure.

Step 2: Find $\{P_{CA}\}$ and $\{p_F\}$

The fixed end actions and equivalent nodal loads are shown in Figure 4.17c.

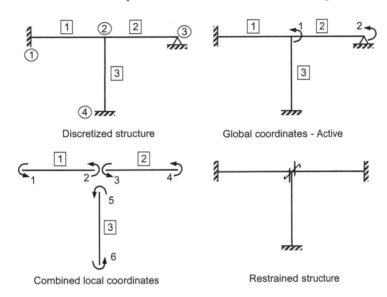

FIGURE 4.17b
Discretized structure, coordinates, and restrained structure.

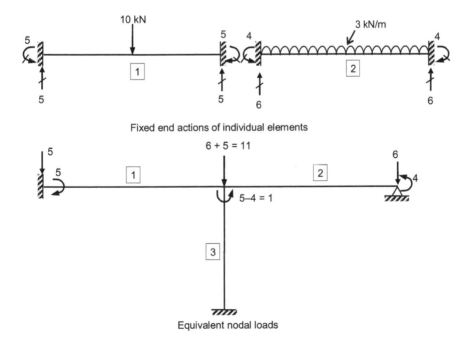

FIGURE 4.17c
Fixed end actions and equivalent nodal loads.

$$\{P_{CA}\} = \left\{ \begin{array}{c} P_{C1} \\ P_{C2} \end{array} \right\} = \left\{ \begin{array}{c} 1 \\ 4 \end{array} \right\} \text{kNm}$$

$$\{p_F\} = \left\{ \begin{array}{c} 5 \\ -5 \\ 4 \\ -4 \\ 0 \\ 0 \end{array} \right\} \text{kNm}$$

Step 3: Develop $[T_{DA}]$
From Figure 4.17d,

$$[T_{DA}] = \begin{array}{c} {\scriptstyle D_1=1 \quad D_2=1} \\ \left[\begin{array}{c:c} 0 & 0 \\ 1 & 0 \\ 1 & 0 \\ 0 & 1 \\ 1 & 0 \\ 0 & 0 \end{array} \right] \end{array}$$

Step 4: Develop $[K_{AA}]$

- $[k^i]$

$$\left[k^1\right] = \left[k^2\right] = \left[k^3\right] = \frac{EI}{4} \left[\begin{array}{cc} 4 & 2 \\ 2 & 4 \end{array} \right] = EI \left[\begin{array}{cc} 1 & 0.5 \\ 0.5 & 1 \end{array} \right]$$

- $[k]$

$$[k] = \left[\begin{array}{ccc} \left[k^1\right] & [0] & [0] \\ {[0]} & \left[k^2\right] & [0] \\ {[0]} & [0] & \left[k^3\right] \end{array} \right] = EI \left[\begin{array}{cccccc} 1 & 0.5 & 0 & 0 & 0 & 0 \\ 0.5 & 1 & 0 & 0 & 0 & 0 \\ 0 & 0 & 1 & 0.5 & 0 & 0 \\ 0 & 0 & 0.5 & 1 & 0 & 0 \\ 0 & 0 & 0 & 0 & 1 & 0.5 \\ 0 & 0 & 0 & 0 & 0.5 & 1 \end{array} \right]$$

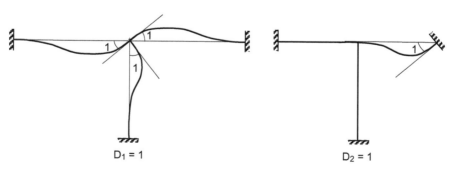

FIGURE 4.17d
Deformed shape of the restrained structure.

- $[K_{AA}]$

$$[K_{AA}]=[T_{DA}]^T[k][T_{DA}]=EI\begin{bmatrix} 3 & 0.5 \\ 0.5 & 1 \end{bmatrix}$$

Step 5: Find $\{D_A\}$

$$\{D_A\}=[K_{AA}]^{-1}\{P_{CA}\}$$

$$\{D_A\}=\begin{Bmatrix} D_1 \\ D_2 \end{Bmatrix}=\frac{1}{EI}\begin{Bmatrix} -0.3636 \\ 4.1818 \end{Bmatrix}$$

Step 6: Find $\{p\}$

$$\{p\}=\{p_F\}+[k][T_{DA}]\{D_A\}=\begin{Bmatrix} 4.818 \\ -5.364 \\ 5.727 \\ 0 \\ -0.364 \\ -0.182 \end{Bmatrix}\text{kNm}$$

Example 4.11

Analyze the frame shown in Figure 4.18a.

Solution

Step 1: Develop the discretized structure (Figure 4.18b)
  Step 2: Find $\{P_{CA}\}$ and $\{p_F\}$ (Figure 4.18c)

$$\{P_{CA}\}=\begin{Bmatrix} P_{C1} \\ P_{C2} \\ P_{C3} \end{Bmatrix}=\begin{Bmatrix} 4\text{ kN} \\ 0\text{ kNm} \\ 3\text{ kNm} \end{Bmatrix}$$

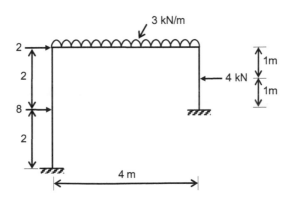

FIGURE 4.18a
Example 4.11.

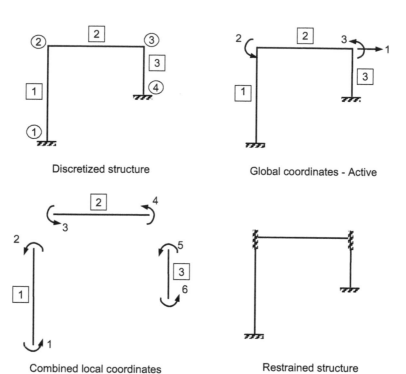

FIGURE 4.18b
Discretized structure, coordinates, and restrained structure.

$$\{p_F\} = \begin{Bmatrix} 4 \\ -4 \\ 4 \\ -4 \\ 1 \\ -1 \end{Bmatrix} \text{kNm}$$

Step 3: Develop $[T_{DA}]$
From Figure 4.18d,

$$[T_{DA}] = \begin{bmatrix} 0.25 & 0 & 0 \\ 0.25 & 1 & 0 \\ 0 & 1 & 0 \\ 0 & 0 & 1 \\ 0.5 & 0 & 1 \\ 0.5 & 0 & 0 \end{bmatrix}$$

with column headings $D_1=1$, $D_2=1$, $D_3=1$.

Step 4: Develop $[K_{AA}]$

- $[k^i]$

$$[k^1] = [k^2] = \frac{EI}{4} \begin{bmatrix} 4 & 2 \\ 2 & 4 \end{bmatrix} = EI \begin{bmatrix} 1 & 0.5 \\ 0.5 & 1 \end{bmatrix}$$

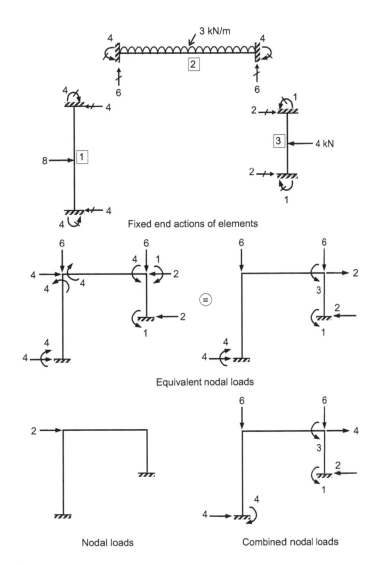

FIGURE 4.18c
Fixed end actions and nodal loads.

$$\left[k^3\right] = \frac{EI}{2}\begin{bmatrix} 4 & 2 \\ 2 & 4 \end{bmatrix} = EI\begin{bmatrix} 2 & 1 \\ 1 & 2 \end{bmatrix}$$

- $[k]$

$$[k] = \begin{bmatrix} \left[k^1\right] & [0] & [0] \\ [0] & \left[k^2\right] & [0] \\ [0] & [0] & \left[k^3\right] \end{bmatrix} = EI\begin{bmatrix} 1 & 0.5 & 0 & 0 & 0 & 0 \\ 0.5 & 1 & 0 & 0 & 0 & 0 \\ 0 & 0 & 1 & 0.5 & 0 & 0 \\ 0 & 0 & 0.5 & 1 & 0 & 0 \\ 0 & 0 & 0 & 0 & 2 & 1 \\ 0 & 0 & 0 & 0 & 1 & 2 \end{bmatrix}$$

- $[K_{AA}]$

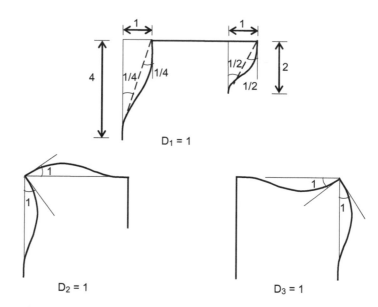

FIGURE 4.18d
Deformed shape of the restrained structure.

$$[K_{AA}] = [T_{DA}]^T [k][T_{DA}] = EI \begin{bmatrix} 1.6875 & 0.375 & 1.5 \\ 0.375 & 2 & 0.5 \\ 1.5 & 0.5 & 3 \end{bmatrix}$$

Step 5: Find $\{D_A\}$

$$\{D_A\} = [K_{AA}]^{-1}\{P_{CA}\}$$

$$\{D_A\} = \left\{ \begin{array}{c} D_1 \\ D_2 \end{array} \right\} = \frac{1}{EI} \left\{ \begin{array}{c} 2.7251 \\ -0.4386 \\ -0.2895 \end{array} \right\}$$

Step 6: Find $\{p\}$

$$\{p\} = \{p_F\} + [k][T_{DA}]\{D_A\} = \left\{ \begin{array}{c} 4.803 \\ -3.417 \\ 3.417 \\ -4.509 \\ 4.509 \\ 2.798 \end{array} \right\} \text{kNm}$$

Example 4.12

Analyze the plane frame shown in Figure 4.19a.

Solution

Step 1: Develop the discretized structure (Figure 4.19b)
Step 2: Define $\{P_A\}$

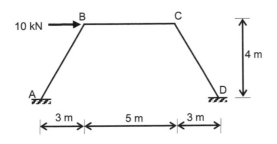

FIGURE 4.19a
Example 4.12.

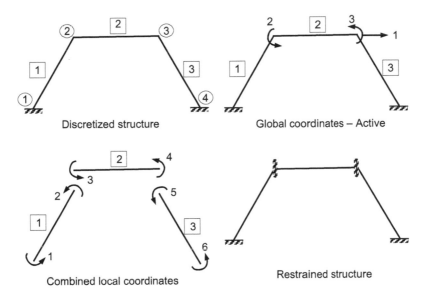

FIGURE 4.19b
Discretized structure, coordinates, and restrained structure.

$$\{P_A\} = \left\{\begin{array}{c} P_1 \\ P_2 \\ P_3 \end{array}\right\} = \left\{\begin{array}{c} 10\,\text{kN} \\ 0\,\text{kNm} \\ 0\,\text{kNm} \end{array}\right\}$$

Step 3: Find $[T_{DA}]$ (Figure 4.19c)
From the deformed shape of the structure corresponding to $D_1 = 1$,

- $d_1 = BB'/l$. But $BB' = BB''/\cos(90-\theta) = 1/\sin\theta = 1.25$

$$\Rightarrow \quad d_1 = 1.25/5 = 0.25$$

- $d_2 = 0.25$
- $d_3 = -(B'B'' + C'C'')/l$

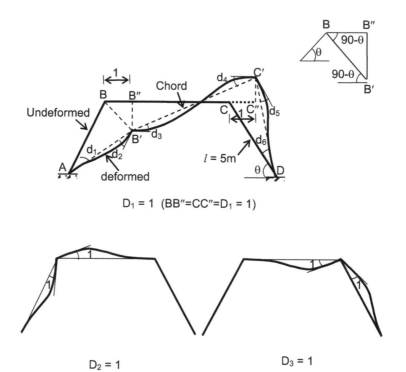

$D_1 = 1$ $(BB''=CC''=D_1 = 1)$

$D_2 = 1$ $D_3 = 1$

FIGURE 4.19c
Deformed shape of the structure.

From the figure, $B'B'' = BB'' \times \tan(90-\theta) = 1 \times \tan(90-\theta)=0.75$

$$C'C'' = B'B'' = 0.75$$

$$\Rightarrow d_3 = -2 \times 0.75/5 = -0.3$$

- $d_4 = -0.3$
- $d_5 = 0.25$
- $d_6 = 0.25$

The other elements of $[T_{DA}]$ are obtained from the deformed shape of the structure corresponding to $D_2 = 1$ and $D_3 = 1$.
     Thus,

$$[T_{DA}]=\begin{array}{ccc} {\scriptstyle D_1=1} & {\scriptstyle D_2=1} & {\scriptstyle D_3=1} \\ \begin{bmatrix} 0.25 & 0 & 0 \\ 0.25 & 1 & 0 \\ -0.3 & 1 & 0 \\ -0.3 & 0 & 1 \\ 0.25 & 0 & 1 \\ 0.25 & 0 & 0 \end{bmatrix} \end{array}$$

Step 4: Develop $[K_{AA}]$
- $[k^i]$

$$\left[k^1\right]=\left[k^2\right]=\left[k^3\right]=\frac{EI}{5}\begin{bmatrix} 4 & 2 \\ 2 & 4 \end{bmatrix}=EI\begin{bmatrix} 0.8 & 0.4 \\ 0.4 & 0.8 \end{bmatrix}$$

- $[k]$

$$[k]=EI\begin{bmatrix} 0.8 & 0.4 & 0 & 0 & 0 & 0 \\ 0.4 & 0.8 & 0 & 0 & 0 & 0 \\ 0 & 0 & 0.8 & 0.4 & 0 & 0 \\ 0 & 0 & 0.4 & 0.8 & 0 & 0 \\ 0 & 0 & 0 & 0 & 0.8 & 0.4 \\ 0 & 0 & 0 & 0 & 0.4 & 0.8 \end{bmatrix}$$

- $[K_{AA}]$

$$[K_{AA}]=[T_{DA}]^T[k][T_{DA}]=EI\begin{bmatrix} 0.516 & -0.06 & -0.06 \\ -0.06 & 1.6 & 0.4 \\ -0.06 & 0.4 & 1.6 \end{bmatrix}$$

Step 5: Find $\{D_A\}$

$$\{D_A\}=[K_{AA}]^{-1}\{P_A\}$$

$$\{D_A\}=\begin{Bmatrix} D_1 \\ D_2 \\ D_3 \end{Bmatrix}=\frac{1}{EI}\begin{Bmatrix} 19.516 \\ 0.586 \\ 0.586 \end{Bmatrix}$$

Step 6: Find $\{p\}$

$$\{p\}=[k][T_{DA}]\{D_A\}=\begin{Bmatrix} 6.089 \\ 6.323 \\ -6.323 \\ -6.323 \\ 6.323 \\ 6.089 \end{Bmatrix}\text{kNm}$$

Example 4.13

Analyze the truss shown in Figure 4.20a.

Solution

Step 1: Develop the discretized structure (Figure 4.20b)
  Step 2: Define $\{P_A\}$

$$\{P_A\}=\begin{Bmatrix} P_1 \\ P_2 \end{Bmatrix}=\begin{Bmatrix} 10 \\ -50 \end{Bmatrix}\text{kN}$$

Step 3: Find $[T_{DA}]$

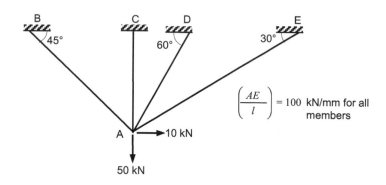

FIGURE 4.20a
Example 4.13.

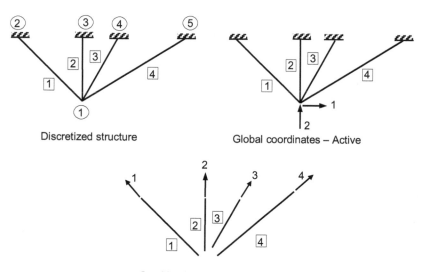

Discretized structure

Global coordinates – Active

Combined local coordinates

FIGURE 4.20b
Discretized structure and coordinates.

- Equilibrium matrix $[e_A]$
  From FBD of Node 1 (Figure 4.20c),

$$\sum F_X = 0 \Rightarrow$$

$$P_1 = p_1 \cos 45 - p_3 \cos 60 - p_4 \cos 30$$

$$\therefore \quad P_1 = 0.707 p_1 + 0 p_2 - 0.5 p_3 - 0.866 p_4 \qquad \text{(a)}$$

$$\sum F_Y = 0 \Rightarrow$$

$$P_2 = -p_1 \sin 45 - p_2 - p_3 \sin 60 - p_4 \sin 30$$

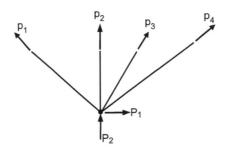

FIGURE 4.20c
FBD of node 1.

$$P_2 = -0.707p_1 - 1p_2 - 0.866p_3 - 0.5p_4 \tag{b}$$

$$\{P_A\} = [e_A]\{p\}$$

From Eqs. (a) and (b),

$$\left\{ \begin{array}{c} P_1 \\ P_2 \end{array} \right\} = \left[\begin{array}{cccc} 0.707 & 0 & -0.5 & -0.866 \\ -0.707 & -1 & -0.866 & -0.5 \end{array} \right] \left\{ \begin{array}{c} p_1 \\ p_2 \\ p_3 \\ p_4 \end{array} \right\}$$

- $[T_{DA}]$

$$[T_{DA}] = [e_A]^T$$

$$= \left[\begin{array}{cc} 0.707 & -0.707 \\ 0 & -1 \\ -0.5 & -0.866 \\ -0.866 & -0.5 \end{array} \right]$$

Step 4: Develop $[K_{AA}]$

- $[k^i]$

$$[k^1] = [k^2] = [k^3] = [k^4] = \frac{AE}{l} = 100 \text{ kN/mm}$$

- $[k]$

$$[k] = \frac{AE}{l} \left[\begin{array}{cccc} 1 & 0 & 0 & 0 \\ 0 & 1 & 0 & 0 \\ 0 & 0 & 1 & 0 \\ 0 & 0 & 0 & 1 \end{array} \right]$$

- $[K_{AA}]$

$$[K_{AA}] = [T_{DA}]^T [k][T_{DA}] = \frac{AE}{l} \begin{bmatrix} 1.5 & 0.366 \\ 0.366 & 2.5 \end{bmatrix}$$

Step 5: Find $\{D_A\}$

$$\{D_A\} = [K_{AA}]^{-1} \{P_A\}$$

$$\{D_A\} = \begin{Bmatrix} D_1 \\ D_2 \end{Bmatrix} = \frac{l}{AE} \begin{Bmatrix} 11.98 \\ -21.76 \end{Bmatrix} = \begin{Bmatrix} 0.1198 \\ -0.2176 \end{Bmatrix} \text{mm}$$

Step 6: Find $\{p\}$

$$\{p\} = [k][T_{DA}]\{D_A\} = \begin{Bmatrix} p_1 \\ p_2 \\ p_3 \\ p_4 \end{Bmatrix} = \begin{Bmatrix} 23.851 \\ 21.756 \\ 12.851 \\ 0.504 \end{Bmatrix} \text{kN}$$

Example 4.14

Find the forces developed in the truss shown in Figure 4.20a due to fabrication errors, if the members AB, AC, and AE were made 5-mm short, 1-mm long, and 2-mm short before the fabrication of truss.

Solution

Step 1: Develop the discretized structure.
   The elements, nodes, and coordinates are shown in Figure 4.20b.
Step 2: Find $\{P_{CA}\}$ and $\{p_F\}$
   The fixed end actions and the nodal loads are shown in Figures 3.30d–h.

$$\{P_{CA}\} = \begin{Bmatrix} P_{C1} \\ P_{C2} \end{Bmatrix} = \begin{Bmatrix} -180.34 \\ 353.55 \end{Bmatrix} \text{kN}$$

$$\{p_F\} = \begin{Bmatrix} 500 \\ -100 \\ 0 \\ 200 \end{Bmatrix} \text{kN}$$

Step 3: Develop $[T_{DA}]$ and $[K_{AA}]$
   $[T_{DA}]$ and $[K_{AA}]$ of the structures are developed in Example 4.13.
Step 4: Find $\{D_A\}$

$$\{D_A\} = [K_{AA}]^{-1} \{P_{CA}\}$$

$$\{D_A\} = \begin{Bmatrix} D_1 \\ D_2 \end{Bmatrix} = \begin{Bmatrix} -1.6051 \\ 1.6494 \end{Bmatrix} \text{mm}$$

Step 5: Find $\{p\}$

$$\{p\} = \{p_F\} + [k][T_{DA}]\{D_A\} = \begin{Bmatrix} 269.906 \\ -264.941 \\ -62.584 \\ 256.531 \end{Bmatrix} kN$$

Problem

4.1 Analyze the following structures using the stiffness method (Figure 4.21).

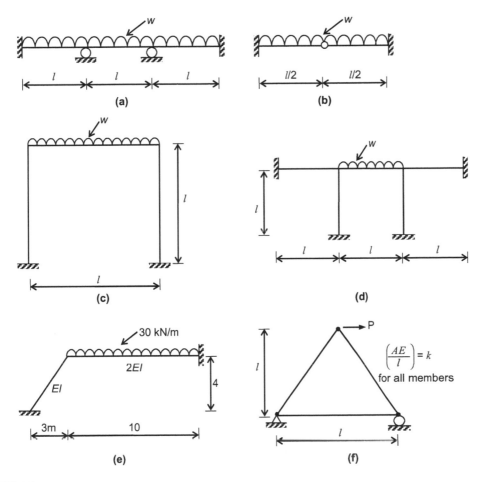

FIGURE 4.21
Problem 4.1.

5

Direct Stiffness Method

5.1 Introduction

In the previous chapter, the displacement transformation matrix $[T_D]$ was used to develop the structure stiffness matrix $[K]$. This concept is useful to understand the physical meaning of $[T_D]$ and $[K]$. However, when the structure to be analyzed becomes complex or large, development of $[T_D]$ itself will be difficult. Moreover, $[T_D]$ depends on the structure to be analyzed, i.e., $[T_D]$ developed for a structure cannot be used to analyze another structure composed of similar elements[*]. Hence, this procedure is not suitable for automatic calculation. In this chapter, an alternate procedure known as the *direct stiffness method* is introduced which is suitable for computer implementation. In this method, transformation matrix for each element of the structure is developed and they are used to develop $[K]$.

The direct stiffness method can be easily understood by considering the analogy between the procedures used to develop $[K]$ in the direct stiffness method and to calculate the second moment of area of composite areas. For this purpose, the steps required to find the second moment of area about the X-axis (I_{XX}) of the composite area shown in Figure 5.1a are listed below.

i. Divide the composite area into regular plane areas. In this case, the composite area is divided into two areas 1 and 2. The areas are A_1 and A_2, as shown in Figure 5.1b.

ii. Locate the centroid of these areas (C_1 and C_2). The axes x_1– and x_2– shown in Figure 5.1c are the centroidal axes for the areas 1 and 2, respectively.

iii. Calculate the second moment of area of 1 and 2 with respect to its centroidal axis (I_{x1} and I_{x2}).

iv. Use parallel axis theorem to transform I_{x1} and I_{x2} to the second moment of areas about the X-axis (I_{X1} and I_{X2}).

$$I_{X1} = I_{x1} + A_1 h_1^2 \,;\, I_{X2} = I_{x2} + A_2 h_2^2$$

v. Calculate I_{XX} by adding I_{X1} and I_{X2}.

$$I_{XX} = I_{X1} + I_{X2}$$

---

[*] For example, $[T_D]$ for a plane truss cannot be used to analyse another plane truss of different configuration.

255

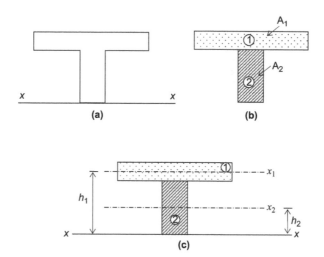

FIGURE 5.1
Steps to find I_{XX} of a composite area: (a) composite area, (b) divide into regular plane areas, and (c) centroid of 1 and 2.

Similar to the above, the different steps involved to develop $[K]$ in the direct stiffness method are as follows:

  i. Divide the structure into number of elements connected to each other by nodes, i.e., develop the discretized structure. This step is similar to divide the composite area into regular areas.

  ii. Define the local axes of the elements and global axes of the structure. The axes x_1, x_2, and X in Figure 5.1c are analogous to local axes and global axes, respectively.

  iii. Develop the stiffness matrix of the elements with respect to the local axes. This is similar in finding I_{x1} and I_{x2}.

  iv. Use the transformation rules to transform the element stiffness from local axes to global axes. This step is similar to use parallel axes theorem to find I_{X1} and I_{X2} using I_{x1} and I_{x2}.

  v. *Assemble the element stiffness matrices in global axes to get the structure stiffness matrix. This is analogous to find I_{XX} by adding I_{X1} and I_{X2}.

The direct stiffness method is explained in detail in this chapter. The important difference between the stiffness method introduced in the previous chapter and the direct stiffness method is that in the stiffness method a single transformation matrix $[T_D]$ is used for the entire structure to find $[K]$, whereas in the direct stiffness method, separate transformation matrix for each element is developed. Both the methods use the same force–displacement relation for the structure, i.e.,

$$\left\{\begin{array}{c} \{P_A\} \\ \hline \{P_R\} \end{array}\right\} = \left[\begin{array}{c|c} [K_{AA}] & [K_{AR}] \\ \hline [K_{RA}] & [K_{RR}] \end{array}\right] = \left\{\begin{array}{c} \{D_A\} \\ \hline \{D_R\} \end{array}\right\}$$

---

* Instead of the term 'addition', 'assemble' is used in the direct stiffness method.

or

$${P} = [K]{D}$$

The two methods differ only in the manner by which $[K]$ is developed.

5.2 Coordinates Used in Direct Stiffness Method

The coordinates used in the direct stiffness method are similar to the coordinates used in the stiffness method (see Section 4.2). The coordinates required for analysis using the direct stiffness method are discussed using the plane frame shown in Figure 5.2a.

The global coordinates* (active and restrained) of the structure are indicated in Figure 5.2b. The load vector $\{P_A\}$, reaction vector $\{P_R\}$, active displacement vector $\{D_A\}$, and restrained displacement vector $\{D_R\}$ of the structure are as follows:

$$\{P_A\} = \begin{Bmatrix} P_1 \\ P_2 \\ P_3 \\ \vdots \\ P_{13} \end{Bmatrix}, \{P_R\} = \begin{Bmatrix} P_{14} \\ P_{15} \\ P_{16} \\ P_{17} \\ P_{18} \end{Bmatrix}, \{D_A\} = \begin{Bmatrix} D_1 \\ D_2 \\ D_3 \\ \vdots \\ D_{13} \end{Bmatrix}, \{D_R\} = \begin{Bmatrix} D_{14} \\ D_{15} \\ D_{16} \\ D_{17} \\ D_{18} \end{Bmatrix}$$

For each element, two types of coordinates are required for analysis: one along the local axes and the other along the global axes. The coordinates for two typical elements (elements 2 and 3) are shown in Figures 5.2c and d. In the figures, the superscript indicates the element number. The force and displacement vectors of the elements are as follows:

- For element 2

$$\{p^2\} = \begin{Bmatrix} p_1^2 \\ p_2^2 \\ p_3^2 \\ p_4^2 \\ p_5^2 \\ p_6^2 \end{Bmatrix}, \{d^2\} = \begin{Bmatrix} d_1^2 \\ d_2^2 \\ d_3^2 \\ d_4^2 \\ d_5^2 \\ d_6^2 \end{Bmatrix}, \{P^2\} = \begin{Bmatrix} P_1^2 \\ P_2^2 \\ P_3^2 \\ P_4^2 \\ P_5^2 \\ P_6^2 \end{Bmatrix}, \{D^2\} = \begin{Bmatrix} D_1^2 \\ D_2^2 \\ D_3^2 \\ D_4^2 \\ D_5^2 \\ D_6^2 \end{Bmatrix}$$

---

* Since the direct stiffness method is meant for implementation using a computer, axial deformations of the structure are considered.

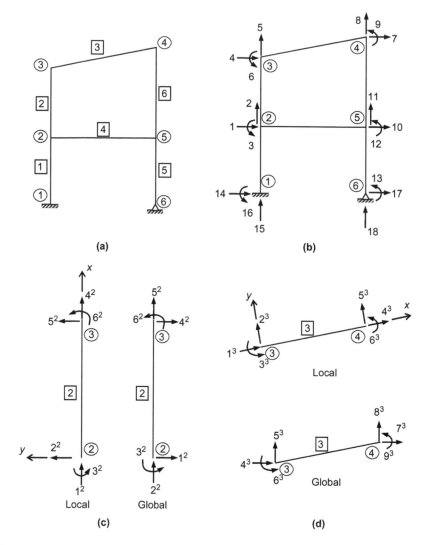

FIGURE 5.2
Coordinates of structure and elements: (a) plane frame with nodes and elements, (b) global coordinates, (c) coordinates of element 2, and (d) coordinates of element 3.

- For element 3

$$\{p^3\} = \begin{Bmatrix} p_1^3 \\ p_2^3 \\ p_3^3 \\ p_4^3 \\ p_5^3 \\ p_6^3 \end{Bmatrix}, \{d^3\} = \begin{Bmatrix} d_1^3 \\ d_2^3 \\ d_3^3 \\ d_4^3 \\ d_5^3 \\ d_6^3 \end{Bmatrix}, \{P^3\} = \begin{Bmatrix} P_1^3 \\ P_2^3 \\ P_3^3 \\ P_4^3 \\ P_5^3 \\ P_6^3 \end{Bmatrix}, \{D^3\} = \begin{Bmatrix} D_1^3 \\ D_2^3 \\ D_3^3 \\ D_4^3 \\ D_5^3 \\ D_6^3 \end{Bmatrix}$$

where $\{p^i\}$ and $\{d^i\}$ are the element force and displacement vectors along the local coordinates. i is the element number. $\{P^i\}$ and $\{D^i\}$ are the element force and displacement vectors along the global coordinates.

The force displacement equations for element i along the local and global axes are

$$\{p^i\} = [k^i]\{d^i\} \tag{5.1}$$

and

$$\{P^i\} = [K^i]\{D^i\} \tag{5.2}$$

where $[k^i]$ and $[K^i]$ are the stiffness matrices of element i with respect to local and global axes.

5.3 Transformation Laws for Vectors

5.3.1 Vector in Plane

Consider a vector $\vec{V}$ (say force or displacement) in a plane. The components of the vector with respect to X- and Y-axes are V_X and V_Y, respectively. New orthogonal axes x and y (Figure 5.3a) are obtained by rotating the X- and Y-axes by an angle θ in the counter clockwise direction. V_x and V_y are the components of $\vec{V}$ with respect to x- and y-axes. The relations between the components of $\vec{V}$ with respect to the old and new axes (Figure 5.3b) are as follows:

$$V_x = CA = CB + BA = V_X \cos\theta + V_Y \sin\theta$$

$$V_y = AD = AE - DE = V_Y \cos\theta - V_X \sin\theta$$

In matrix form,

$$\begin{Bmatrix} V_x \\ V_y \end{Bmatrix} = \begin{bmatrix} \cos\theta & \sin\theta \\ -\sin\theta & \cos\theta \end{bmatrix} \begin{Bmatrix} V_X \\ V_Y \end{Bmatrix}$$

or

$$\begin{Bmatrix} V_x \\ V_y \end{Bmatrix} = [R] \begin{Bmatrix} V_X \\ V_Y \end{Bmatrix} \tag{5.3}$$

where

$$[R] = \begin{bmatrix} \cos\theta & \sin\theta \\ -\sin\theta & \cos\theta \end{bmatrix} = \begin{bmatrix} l_1 & m_1 \\ l_2 & m_2 \end{bmatrix}$$

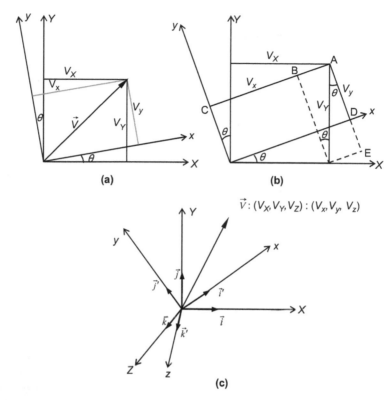

FIGURE 5.3
Transformation of vectors from global to local axes.

is the *rotation transformation matrix*. It can be verified that $[R]$ is a orthogonal matrix (i.e., $[R]^{-1} = [R]^T$). l_1, m_1, and l_2, m_2 are the direction cosines of x and y axes.

Equation 5.3 is the transformation law of vectors in a plane. Using this equation, it is possible to find the components of $\vec{V}$ with respect to local axes (x- and y-axes) if the components of the vector are known with respect to global axes (X- and Y-axes).

5.3.2 Vector in Space

The transformation law for a vector in space is derived in this section. Let $\vec{i}, \vec{j}, \vec{k}$ and $\vec{i}', \vec{j}', \vec{k}'$ are the unit vectors along the global and local axes (Figure 5.3c). The components of the vector with respect to global and local axes are V_X, V_Y, V_Z and V_x, V_y, V_z, respectively. Then,

$$\vec{V} = V_X \vec{i} + V_Y \vec{j} + V_Z \vec{k} \tag{a}$$

Also

$$\vec{V} = V_x \vec{i}' + V_y \vec{j}' + V_z \vec{k}' \tag{b}$$

The components V_x, V_y, and V_z are obtained by taking the dot product of $\vec{V}$ with the unit vectors $\vec{i}', \vec{j}', \vec{k}'$. Thus, using Eq. (a)

$$V_x = \vec{V} \cdot \vec{i}' = \left(V_X \vec{i} + V_Y \vec{j} + V_Z \vec{k} \right) \cdot \vec{i}'$$

$$= \left(\vec{i} \cdot \vec{i}' \right) V_X + \left(\vec{j} \cdot \vec{i}' \right) V_Y + \left(\vec{k} \cdot \vec{i}' \right) V_Z \tag{c}$$

Similarly,

$$V_y = \vec{V} \cdot \vec{j}' = \left(\vec{i} \cdot \vec{j}' \right) V_X + \left(\vec{j} \cdot \vec{j}' \right) V_Y + \left(\vec{k} \cdot \vec{j}' \right) V_Z \tag{d}$$

$$V_z = \vec{V} \cdot \vec{k}' = \left(\vec{i} \cdot \vec{k}' \right) V_X + \left(\vec{j} \cdot \vec{k}' \right) V_Y + \left(\vec{k} \cdot \vec{k}' \right) V_Z \tag{e}$$

Equations (c) to (e) in the matrix form give

$$\begin{Bmatrix} V_x \\ V_y \\ V_z \end{Bmatrix} = \begin{bmatrix} \vec{i} \cdot \vec{i}' & \vec{j} \cdot \vec{i}' & \vec{k} \cdot \vec{i}' \\ \vec{i} \cdot \vec{j}' & \vec{j} \cdot \vec{j}' & \vec{k} \cdot \vec{j}' \\ \vec{i} \cdot \vec{k}' & \vec{j} \cdot \vec{k}' & \vec{k} \cdot \vec{k}' \end{bmatrix} \begin{Bmatrix} V_X \\ V_Y \\ V_Z \end{Bmatrix}$$

$$= \begin{bmatrix} l_1 & m_1 & n_1 \\ l_2 & m_2 & n_2 \\ l_3 & m_3 & n_3 \end{bmatrix} \begin{Bmatrix} V_X \\ V_Y \\ V_Z \end{Bmatrix} = [R] \begin{Bmatrix} V_x \\ V_y \\ V_z \end{Bmatrix} \tag{5.4}$$

where l_1, m_1, n_1; l_2, m_2, n_2; and l_3, m_3, n_3 are the direction cosines of x-, y- and z-axes with respect to the global axes. $[R]$ is the rotation matrix for a space vector, which is again is orthogonal matrix.

Equation (5.4) is the transformation law for vectors in space.

5.4 Element Stiffness Matrix with Respect to Global Axes $[k^i]$

The stiffness matrix of different types of elements with respect to local axes $[k^i]$ was discussed in Chapters 2 and 4. In a structure where many elements with different orientations meet at a node, it becomes necessary to transform the element stiffness matrix along the local axes to the common global axes of the structure so that the structure stiffness matrix can be developed. The different transformation matrices necessary for this purpose are discussed below.

5.4.1 Transformation of End Actions and Displacements

5.4.1.1 *End Transformation Matrix [t^i]*

Consider a plane frame element i with near node N and for node F (Figure 5.4a). The end actions at any node of the element are shown in Figure 5.4b. In the figure, p_a^i, p_b^i, and p_c^i are the axial force, shear force, and bending moment of the element i with respect to the local axes. The axial force p_a^i and shear force p_b^i are added vectorially to get the resultant force

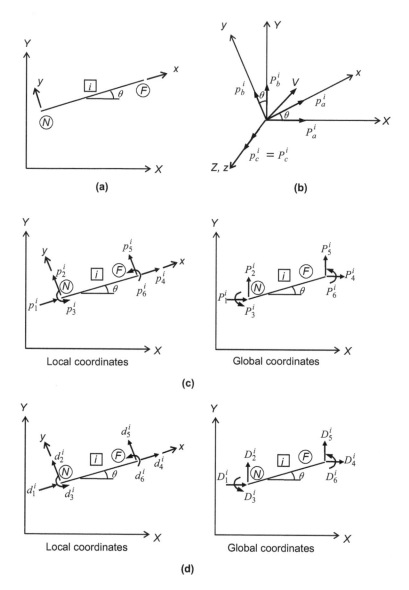

FIGURE 5.4
Transformation of element forces and displacements: (a) plane frame element, (b) forces at a node, (c) end action in a frame element, and (d) end displacements in a plane frame element.

vector $\vec{V}_i$. P_a^i and P_b^i are the components of $\vec{V}_i$ along the global X- and Y-axes. P_c^i is the bending moment at the node along the global axis Z. The forces along the local axes p_a^i, p_b^i are related to the forces along the global axes p_a^i, p_b^i using Eq. 5.3. Thus,

$$p_a^i = l_1 P_a^i + m_1 P_b^i$$

$$p_b^i = l_2 P_a^i + m_2 P_b^i$$

From Figure 5.4b,

$$l_1 = \cos\theta = c, m_1 = \sin\theta = s, l_2 = -s \text{ and } m_2 = c.$$

Hence,

$$p_a^i = cP_a^i + sP_b^i \tag{a}$$

$$p_b^i = -sP_a^i + cP_b^i \tag{b}$$

Since the local z-axis and global Z-axis are the same,

$$p_c^i = P_c^i \tag{c}$$

Equations (a) to (c) in the matrix form give

$$\begin{Bmatrix} p_a^i \\ p_b^i \\ p_c^i \end{Bmatrix} = \begin{bmatrix} c & s & 0 \\ -s & c & 0 \\ 0 & 0 & 1 \end{bmatrix} \begin{Bmatrix} P_a^i \\ P_b^i \\ P_c^i \end{Bmatrix}$$

$$\Rightarrow \quad \begin{Bmatrix} p_a^i \\ p_b^i \\ p_c^i \end{Bmatrix} = \begin{bmatrix} t^i \end{bmatrix} \begin{Bmatrix} P_a^i \\ P_b^i \\ P_c^i \end{Bmatrix} \tag{5.5}$$

where $[t^i]$ is the *end transformation matrix* for the element i.

5.4.1.2 Element Transformation Matrix $[T^i]$

Using Eq. (5.5), the end actions at both the nodes are transformed from global axes to local axes as

$$\begin{Bmatrix} p_1^i \\ p_2^i \\ p_3^i \\ \hdashline p_4^i \\ p_5^i \\ p_6^i \end{Bmatrix} = \left[\begin{array}{ccc:ccc} c & s & 0 & 0 & 0 & 0 \\ -s & c & 0 & 0 & 0 & 0 \\ 0 & 0 & 1 & 0 & 0 & 0 \\ \hdashline 0 & 0 & 0 & c & s & 0 \\ 0 & 0 & 0 & -s & c & 0 \\ 0 & 0 & 0 & 0 & 0 & 1 \end{array} \right] \begin{Bmatrix} P_1^i \\ P_2^i \\ P_3^i \\ \hdashline P_4^i \\ P_5^i \\ P_6^i \end{Bmatrix} \begin{array}{l} \left. \vphantom{\begin{matrix} a\\a\\a \end{matrix}} \right\} \text{Actions at near node} \\ \\ \left. \vphantom{\begin{matrix} a\\a\\a \end{matrix}} \right\} \text{Actions at far node} \end{array}$$

or

$$\{p^i\} = [T^i]\{P^i\} \tag{5.6}$$

where $\{p^i\}$ and $\{P^i\}$ are the element force vectors in local and global coordinates. $[T]$ is the *element transformation matrix* which can be written in terms of $[t^i]$ as

$$[T^i] = \left[\begin{array}{c|c} [t^i] & [0] \\ \hline [0] & [t^i] \end{array}\right] \tag{5.7}$$

The end displacements of the frame element along local and global coordinates are shown in Figure 5.4d. Comparing this with Figure 5.4c indicates that $[T]$ can be also used for transforming end displacements of the element from global to local axes. Hence,

$$\left\{\begin{array}{c} d^i_1 \\ d^i_2 \\ d^i_3 \\ \hline d^i_4 \\ d^i_5 \\ d^i_6 \end{array}\right\} = \left[\begin{array}{ccc|ccc} c & s & 0 & 0 & 0 & 0 \\ -s & c & 0 & 0 & 0 & 0 \\ 0 & 0 & 1 & 0 & 0 & 0 \\ \hline 0 & 0 & 0 & c & s & 0 \\ 0 & 0 & 0 & -s & c & 0 \\ 0 & 0 & 0 & 0 & 0 & 1 \end{array}\right] \left\{\begin{array}{c} D^i_1 \\ D^i_2 \\ D^i_3 \\ \hline D^i_4 \\ D^i_5 \\ D^i_6 \end{array}\right\} \begin{array}{l} \Big\} \text{Displacements of near node} \\ \\ \Big\} \text{Displacements of far node} \end{array}$$

or

$$\{d^i\} = [T^i]\{D^i\} \tag{5.8}$$

where $\{d^i\}$ and $\{D^i\}$ are the end displacement vectors of element i in local and global coordinates.

5.4.1.3 Direction Cosines

Knowing the coordinates of the nodes of the element, the direction cosines are determined as

$$c = \cos\theta = \frac{X_F - X_N}{l} \tag{5.9a}$$

$$s = \sin\theta = \frac{Y_F - Y_N}{l} \tag{5.9b}$$

where l is the length of the element and is equal to

$$l = \sqrt{(X_F - X_N)^2 + (Y_F - Y_N)^2} \tag{5.9c}$$

(X_N, Y_N) and (X_F, Y_F) are the coordinates of the near and far nodes of the element (Figure 5.5). The use of the above equations automatically takes care of the sign of the direction cosines for different orientations of the element.

5.4.1.4 Principle of Contragradience

$\{P^i\}$ and $\{p^i\}$ are two statically equivalent force vectors and $\{D^i\}$, $\{d^i\}$ are their corresponding displacements. Then according to the principle of contragradience (proof is given in

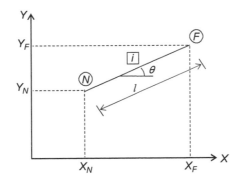

FIGURE 5.5
Coordinates of the nodes of the element.

Section 4.4.2), if one transformation is known, for example, from global displacements $\{D^i\}$ to local displacements $\{d^i\}$, then the force transformation from local ($\{p^i\}$) to global ($\{P^i\}$) coordinates will be equal to the transpose of the displacement transformation, provided both force and displacement vectors are conjugate vectors. Conjugate means that the force and displacement vectors only produce work in the direction of displacement.

From Eq. 5.8,

$$\{d^i\} = [T^i]\{D^i\},$$

then as per the principle of contragradience,

$$\{P^i\} = [T^i]^T\{p^i\} \tag{5.10}$$

5.4.2 Transformation of Element Stiffness Matrix

From Eqs. 5.1 and 5.2,

$$\{p^i\} = [k^i]\{d^i\} \tag{a}$$

$$\{P^i\} = [K^i]\{D^i\} \tag{b}$$

Substituting Eq. (a) in Eq. (5.10) gives

$$\{P^i\} = [T^i]^T\{p^i\} = [T^i]^T[k^i]\{d^i\}$$

Using Eq. (5.8)

$$= [T^i]^T[k^i][T^i]\{D^i\}$$

But

$$\{P^i\} = [K^i]\{D^i\}$$

Hence,

$$\left[K^i\right]=\left[T^i\right]^T\left[k^i\right]\left[T^i\right] \tag{5.11}$$

The transformation matrices ($[t^i]$ and $[T^i]$) and element stiffness matrices of typical elements are given below. The local and global coordinates of these elements are shown in Figure 5.6.

5.4.2.1 Transformation Matrices and Element Stiffness Matrix

a. Plane truss element* (Figure 5.6a)
- End transformation matrix $[t^i]$

$$[t^i]=\begin{bmatrix} l_1 & m_1 \end{bmatrix}=\begin{bmatrix} c & s \end{bmatrix} \tag{5.12a}$$

- Element transformation matrix $[T^i]$

$$[T^i]=\left[\begin{array}{cc|cc} l_1 & m_1 & 0 & 0 \\ \hline 0 & 0 & l_1 & m_1 \end{array}\right]=\left[\begin{array}{cc|cc} c & s & 0 & 0 \\ \hline 0 & 0 & c & s \end{array}\right] \tag{5.12b}$$

- Element stiffness matrix in local coordinates $[k^i]$

$$[k^i]=\frac{AE}{l}\begin{bmatrix} 1 & -1 \\ -1 & 1 \end{bmatrix} \tag{5.12c}$$

- Element stiffness matrix in global coordinates $[K^i]$

$$\left[K^i\right]=\left[T^i\right]^T\left\{k^i\right\}\left[T^i\right]$$

$$=\frac{AE}{l}\begin{bmatrix} c^2 & cs & -c^2 & -cs \\ cs & s^2 & -cs & -s^2 \\ -c^2 & -cs & c^2 & cs \\ -cs & -s^2 & cs & s^2 \end{bmatrix} \tag{5.12d}$$

b. Space truss element (Figure 5.6b)
- End transformation matrix $[t^i]$

$$[t^i]=\begin{bmatrix} l_1 & m_1 & n_1 \end{bmatrix} \tag{5.13a}$$

where

$$l_1=\frac{X_F-X_N}{l}=\frac{(X_F-X_N)}{\sqrt{(X_F-X_N)^2+(Y_F-Y_N)^2+(Z_F-Z_N)^2}}$$

---

* $[k^i]$ of a plane truss element is derived from first principles in Example 2.2.

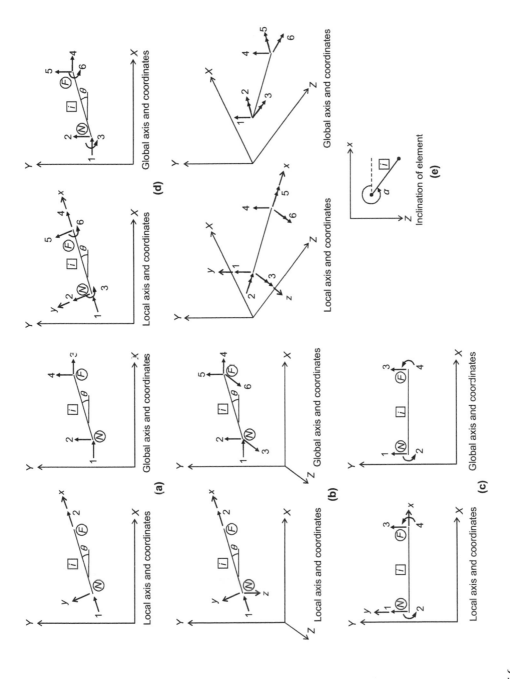

FIGURE 5.6
Local and global coordinates; axes of elements: (a) plane truss element, (b) space truss element, (c) beam element, (d) plane frame element, (e) grid element, and (f) space frame element.

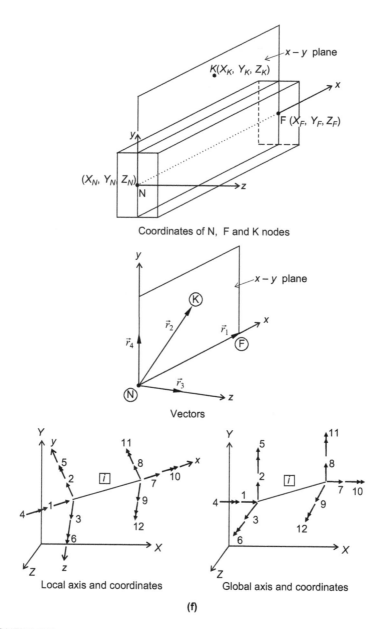

Coordinates of N, F and K nodes

Vectors

Local axis and coordinates   Global axis and coordinates

(f)

FIGURE 5.6 (CONTINUED)
Local and global coordinates; axes of elements: (a) plane truss element, (b) space truss element, (c) beam element, (d) plane frame element, (e) grid element, and (f) space frame element.

$$m_1 = \frac{Y_F - Y_N}{l} = \frac{(Y_F - Y_N)}{\sqrt{(X_F - X_N)^2 + (Y_F - Y_N)^2 + (Z_F - Z_N)^2}}$$

$$n_1 = \frac{Z_F - Z_N}{l} = \frac{(Z_F - Z_N)}{\sqrt{(X_F - X_N)^2 + (Y_F - Y_N)^2 + (Z_F - Z_N)^2}}$$

- Element transformation matrix $[T^i]$

$$[T^i] = \left[\begin{array}{ccc:ccc} l_1 & m_1 & n_1 & 0 & 0 & 0 \\ \hdashline 0 & 0 & 0 & l_1 & m_1 & n_1 \end{array}\right] \qquad (5.13b)$$

- Element stiffness matrix in local coordinates $[k^i]$

$$[k^i] = \frac{AE}{l}\left[\begin{array}{cc} 1 & -1 \\ -1 & 1 \end{array}\right] \qquad (5.13c)$$

- Element stiffness matrix in global coordinates $[K^i]$

$$[K^i] = \frac{AE}{l}\left[\begin{array}{cccccc} l_1^2 & l_1 m_1 & l_1 n_1 & -l_1^2 & -l_1 m_1 & -l_1 n_1 \\ m_1 l_1 & m_1^2 & m_1 n_1 & -m_1 l_1 & -m_1^2 & -m_1 n_1 \\ n_1 l_1 & n_1 m_1 & n_1^2 & -n_1 l_1 & -n_1 m_1 & -n_1^2 \\ -l_1^2 & -l_1 m_1 & -l_1 n_1 & l_1^2 & l_1 m_1 & l_1 n_1 \\ -m_1 l_1 & -m_1^2 & -m_1 n_1 & m_1 l_1 & m_1^2 & m_1 n_1 \\ -n_1 l_1 & -n_1 m_1 & -n_1^2 & n_1 l_1 & n_1 m_1 & n_1^2 \end{array}\right] \qquad (5.13d)$$

c. Beam element (Figure 5.6c)
  - End transformation matrix $[t^i]$

$$[t^i] = \left[\begin{array}{cc} 1 & 0 \\ 0 & 1 \end{array}\right] \qquad (5.14a)$$

- Element transformation matrix $[T^i]$

$$[T^i] = \left[\begin{array}{cc:cc} 1 & 0 & 0 & 0 \\ 0 & 1 & 0 & 0 \\ \hdashline 0 & 0 & 1 & 0 \\ 0 & 0 & 0 & 1 \end{array}\right] \qquad (5.14b)$$

- Element stiffness matrix in local and global coordinates

$$[k^i] = [K^i] = \left[\begin{array}{cccc} \dfrac{12EI}{l^3} & \dfrac{6EI}{l^2} & -\dfrac{12EI}{l^3} & \dfrac{6EI}{l^2} \\[2mm] \dfrac{6EI}{l^2} & \dfrac{4EI}{l} & -\dfrac{6EI}{l^2} & \dfrac{2EI}{l} \\[2mm] -\dfrac{12EI}{l^3} & -\dfrac{6EI}{l^2} & \dfrac{12EI}{l^3} & -\dfrac{6EI}{l^2} \\[2mm] \dfrac{6EI}{l^2} & \dfrac{2EI}{l} & -\dfrac{6EI}{l^2} & \dfrac{4EI}{l} \end{array}\right] \qquad (5.14c)$$

d. Plane frame element (Figure 5.6d)
  • End transformation matrix $[t^i]$

$$[t^i] = \begin{bmatrix} l_1 & m_1 & 0 \\ l_2 & m_2 & 0 \\ 0 & 0 & 1 \end{bmatrix} = \begin{bmatrix} c & s & 0 \\ -s & c & 0 \\ 0 & 0 & 1 \end{bmatrix} \qquad (5.15a)$$

  • Element transformation matrix $[T^i]$

$$[T^i] = \begin{bmatrix} c & s & 0 & 0 & 0 & 0 \\ -s & c & 0 & 0 & 0 & 0 \\ 0 & 0 & 1 & 0 & 0 & 0 \\ 0 & 0 & 0 & c & s & 0 \\ 0 & 0 & 0 & -s & c & 0 \\ 0 & 0 & 0 & 0 & 0 & 1 \end{bmatrix} \qquad (5.15b)$$

  • Element stiffness matrix in local coordinates

$$[k^i] = \begin{bmatrix} z & 0 & 0 & -z & 0 & 0 \\ 0 & k & m & 0 & -k & m \\ 0 & m & a & 0 & -m & b \\ -z & 0 & 0 & z & 0 & 0 \\ 0 & -k & -m & 0 & k & -m \\ 0 & m & b & 0 & -m & a \end{bmatrix} \qquad (5.15c)$$

where $z = AE/l$, $k = 12EI/l^3$, $a = 4EI/l$, $m = 6EI/l^2$, and $b = 2EI/l$
  • Element stiffness matrix in global coordinates

$$[K^i] = \begin{bmatrix} f & g & h & -f & -g & h \\ g & p & q & -g & -p & q \\ h & q & a & -h & -q & b \\ -f & -g & -h & f & g & -h \\ -g & -p & -q & g & p & -q \\ h & q & b & -h & -q & a \end{bmatrix} \qquad (5.15d)$$

where $f = zc^2 + ks^2$, $h = -ms$, $g = (z - k)cs$, $q = mc$, and $p = zs^2 + kc^2$

e. Grid element (Figure 5.6e)
  Figure 5.6e shows a grid element lying in X–Z plane and the deflection of the element occurs along the Y-axis.
  • End transformation matrix $[t^i]$

$$[t^i] = \begin{bmatrix} 1 & 0 & 0 \\ 0 & \cos\alpha & -\sin\alpha \\ 0 & \sin\alpha & \cos\alpha \end{bmatrix} \qquad (5.16a)$$

where

$$\cos\alpha = \frac{X_F - X_N}{l}, \sin\alpha = \frac{-(Z_F - Z_N)}{l}, l = \sqrt{(X_F - X_N)^2 + (Z_F - Z_N)^2}$$

- Element transformation matrix $[T^i]$

$$[T^i] = \begin{bmatrix} 1 & 0 & 0 & 0 & 0 & 0 \\ 0 & \cos\alpha & -\sin\alpha & 0 & 0 & 0 \\ 0 & \sin\alpha & \cos\alpha & 0 & 0 & 0 \\ 0 & 0 & 0 & 1 & 0 & 0 \\ 0 & 0 & 0 & 0 & \cos\alpha & -\sin\alpha \\ 0 & 0 & 0 & 0 & \sin\alpha & \cos\alpha \end{bmatrix} \tag{5.16b}$$

- Element stiffness matrix in local coordinates

$$[k^i] = \begin{bmatrix} \frac{12EI}{l^3} & 0 & \frac{6EI}{l^2} & -\frac{12EI}{l^3} & 0 & \frac{6EI}{l^2} \\ 0 & \frac{GJ}{l} & 0 & 0 & -\frac{GJ}{l} & 0 \\ \frac{6EI}{l^2} & 0 & \frac{4EI}{l} & -\frac{6EI}{l^2} & 0 & \frac{2EI}{l} \\ -\frac{12EI}{l^3} & 0 & -\frac{6EI}{l^2} & \frac{12EI}{l^3} & 0 & -\frac{6EI}{l^2} \\ 0 & -\frac{GJ}{l} & 0 & 0 & \frac{GJ}{l} & 0 \\ \frac{6EI}{l^2} & 0 & \frac{2EI}{l} & -\frac{6EI}{l^2} & 0 & \frac{4EI}{l} \end{bmatrix} \tag{5.16c}$$

f. Space frame element (Figure 5.6f)

- Direction cosines of local axes

The centroidal axis, minor principal axis of the section, and major principal axis of the sections are taken as the local x-, y-, and z-axis of the element. The direction cosines of the local axes are calculated using the coordinates of three points. Out of these points, two points are the near and far nodes (N and F) of the element. The third node known as the K node is any point lying in the x–y plane (Figure 5.6f).

The vectors $\vec{r}_1$ and $\vec{r}_2$ are the displacement vectors directed from N to F and N to K, respectively. $\vec{r}_3$ and $\vec{r}_4$ are vectors along the z- and y-axes. The components of the vectors $\vec{r}_1$ and $\vec{r}_2$ are

$$\vec{r}_1 = (X_F - X_N)\vec{i} + (Y_F - Y_N)\vec{j} + (Z_F - Z_N)\vec{k} = X_x\vec{i} + X_y\vec{j} + X_z\vec{k}$$

$$\vec{r}_2 = (X_K - X_N)\vec{i} + (Y_K - Y_N)\vec{j} + (Z_K - Z_N)\vec{k} = K_x\vec{i} + K_y\vec{j} + K_z\vec{k}$$

where $\vec{i}, \vec{j}$ and $\vec{k}$ are the unit vectors along the global X-, Y-, and Z- axes.

Since the vectors $\vec{r}_1$ and $\vec{r}_2$ are the x–y plane, a vector normal to these vectors will be in the direction of z-axis. Hence, $\vec{r}_3$ is obtained as the cross product of $\vec{r}_1$ and $\vec{r}_2$.

$$\vec{r}_3 = \vec{r}_1 \times \vec{r}_2 = Z_x \vec{i} + Z_y \vec{j} + Z_z \vec{k}$$

where $Z_x = X_y K_z - X_z K_y$, $Z_y = X_z K_x - X_x K_z$, $Z_z = X_x K_y - X_y K_x$

The vector $\vec{r}_4$ is obtained as the cross product of vectors $\vec{r}_3$ and $\vec{r}_1$. Hence,

$$\vec{r}_4 = \vec{r}_3 \times \vec{r}_1 = Y_x \vec{i} + Y_y \vec{j} + Y_z \vec{k}$$

where $Y_x = Z_y X_z - Z_z X_y$, $Y_y = Z_z X_x - Z_x X_z$, $Y_z = Z_x X_y - Z_y X_x$

The components of the unit vectors in the direction of $\vec{r}_1$, $\vec{r}_4$, and $\vec{r}_3$ are the direction cosines of the local axes of the space frame element.

The direction cosines of the x-axis are

$$l_1 = \frac{X_x}{X}, \, m_1 = \frac{X_y}{X}, \, \text{and } n_1 = \frac{X_z}{X} \tag{5.17a}$$

where $X = \sqrt{X_x^2 + X_y^2 + X_z^2}$

The direction cosines of the y-axis are

$$l_2 = \frac{Y_x}{Y}, \, m_2 = \frac{Y_y}{Y}, \, n_2 = \frac{Y_z}{Y} \tag{5.17b}$$

where $Y = \sqrt{Y_x^2 + Y_y^2 + Y_z^2}$

The direction cosines of the z-axis are

$$l_3 = \frac{Z_x}{Z}, \, m_3 = \frac{Z_y}{Z}, \, n_3 = \frac{Z_z}{Z} \tag{5.17c}$$

where $Z = \sqrt{Z_x^2 + Z_y^2 + Z_z^2}$

- End transformation matrix $[t^i]$

$$[t^i] = \begin{bmatrix} l_1 & m_1 & n_1 & 0 & 0 & 0 \\ l_2 & m_2 & n_2 & 0 & 0 & 0 \\ l_3 & m_3 & n_3 & 0 & 0 & 0 \\ 0 & 0 & 0 & l_1 & m_1 & n_1 \\ 0 & 0 & 0 & l_2 & m_2 & n_2 \\ 0 & 0 & 0 & l_3 & m_3 & n_3 \end{bmatrix} \tag{5.17d}$$

- Element transformation matrix $[T^i]$

$$[T^i] = \begin{bmatrix}
l_1 & m_1 & n_1 & 0 & 0 & 0 & 0 & 0 & 0 & 0 & 0 & 0 \\
l_2 & m_2 & n_2 & 0 & 0 & 0 & 0 & 0 & 0 & 0 & 0 & 0 \\
l_3 & m_3 & n_3 & 0 & 0 & 0 & 0 & 0 & 0 & 0 & 0 & 0 \\
0 & 0 & 0 & l_1 & m_1 & n_1 & 0 & 0 & 0 & 0 & 0 & 0 \\
0 & 0 & 0 & l_2 & m_2 & n_2 & 0 & 0 & 0 & 0 & 0 & 0 \\
0 & 0 & 0 & l_3 & m_3 & n_3 & 0 & 0 & 0 & 0 & 0 & 0 \\
0 & 0 & 0 & 0 & 0 & 0 & l_1 & m_1 & n_1 & 0 & 0 & 0 \\
0 & 0 & 0 & 0 & 0 & 0 & l_2 & m_2 & n_2 & 0 & 0 & 0 \\
0 & 0 & 0 & 0 & 0 & 0 & l_3 & m_3 & n_3 & 0 & 0 & 0 \\
0 & 0 & 0 & 0 & 0 & 0 & 0 & 0 & 0 & l_1 & m_1 & n_1 \\
0 & 0 & 0 & 0 & 0 & 0 & 0 & 0 & 0 & l_2 & m_2 & n_2 \\
0 & 0 & 0 & 0 & 0 & 0 & 0 & 0 & 0 & l_3 & m_3 & n_3
\end{bmatrix} \quad (5.17e)$$

- Element stiffness matrix in local coordinates

$$[K] = \begin{bmatrix}
EA/l & 0 & 0 & 0 & 0 & 0 \\
0 & 12EI_Z/l^3 & 0 & 0 & 0 & 6EI_Z/l^2 \\
0 & 0 & 12EI_Y/l^3 & 0 & -6EI_Y/l^2 & 0 \\
0 & 0 & 0 & GJ/l & 0 & 0 \\
0 & 0 & -6EI_Y/l^2 & 0 & 4EI_Y/l & 0 \\
0 & 6EI_Z/l^2 & 0 & 0 & 0 & 4EI_Z/l \\
-EA/l & 0 & 0 & 0 & 0 & 0 \\
0 & -12EI_Z/l^3 & 0 & 0 & 0 & -6EI_Z/l^2 \\
0 & 0 & -12EI_Y/l^3 & 0 & 6EI_Y/l^2 & 0 \\
0 & 0 & 0 & -GJ/l & 0 & 0 \\
0 & 0 & -6EI_Y/l^2 & 0 & 2EI_Y/l & 0 \\
0 & 6EI_Z/l^2 & 0 & 0 & 0 & 2EI_Z/l
\end{bmatrix}$$

$$\begin{bmatrix}
-EA/l & 0 & 0 & 0 & 0 & 0 \\
0 & -12EI_Z/l^3 & 0 & 0 & 0 & 6EI_Z/l^2 \\
0 & 0 & -12EI_Y/l^3 & 0 & -6EI_Y/l^2 & 0 \\
0 & 0 & 0 & -GJ/l & 0 & 0 \\
0 & 0 & 6EI_Y/l^2 & 0 & 2EI_Y/l & 0 \\
0 & -6EI_Z/l^2 & 0 & 0 & 0 & 2EI_Z/l \\
EA/l & 0 & 0 & 0 & 0 & 0 \\
0 & 12EI_Z/l^3 & 0 & 0 & 0 & -6EI_Z/l^2 \\
0 & 0 & 12EI_Y/l^3 & 0 & 6EI_Y/l^2 & 0 \\
0 & 0 & 0 & GJ/l & 0 & 0 \\
0 & 0 & 6EI_Y/l^2 & 0 & 4EI_Y/l & 0 \\
0 & -6EI_Z/l^2 & 0 & 0 & 0 & 4EI_Z/l
\end{bmatrix} \quad (5.17f)$$

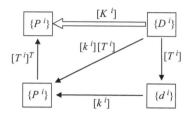

FIGURE 5.7
Transformation between local and global coordinates of an element.

5.5 Transformations Used in Direct Stiffness Method

Figure 5.7 shows the various transformations between the local and global coordinates of an element.

From Figure 5.7,

$$\{d^i\} = [T^i]\{D^i\}$$

$$\{p^i\} = [k^i]\{d^i\} = [k^i][T^i]\{D\}^i \qquad (5.18)$$

$$\{P^i\} = [T^i]^T\{p^i\} = [T^i]^T[k^i][T^i]\{D^i\} = [K^i]\{D^i\}$$

Thus,

$$[K^i] = [T^i]^T[k^i][T^i]$$

5.6 Assembly of Equations and Structure Stiffness Matrix

The force–displacement equations of the elements in global coordinates are assembled to get the force–displacement equations of the structure. The element equations are assembled using compatibility conditions and equilibrium equations for the nodes of the structure. The compatibility condition requires that the displacements of the elements meeting at a node should be same. The equilibrium equation demands that at the node, the forces from the element must be equal to the external forces acting at the node. After assembling the equations, the structure stiffness matrix is obtained.

The process of assemblage is illustrated by considering a truss shown in Figure 5.8a. The forces and displacements of the structure and elements with respect to the global axes are indicated in Figures 5.8b and c.

The force–displacement relations of the elements in global coordinates are as follows.

Element 1

$$\{P^1\} = [K^1]\{D^1\}$$

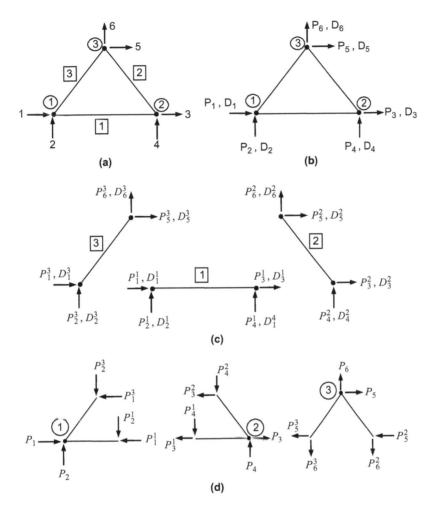

FIGURE 5.8
Assembly of elements—plane truss. (a) Truss and its global coordinates. (b) Forces and displacements of the structure. (c) Forces and displacements of the element in global coordinates. (d) FBD of nodes.

or

$$
\left\{ \begin{array}{c} P_1^1 \\ P_2^1 \\ P_3^1 \\ P_4^1 \end{array} \right\} = \begin{array}{c} \\ \begin{array}{cccc} 1 & 2 & 3 & 4 \end{array} \\ \left[\begin{array}{cccc} a_1 & a_2 & a_3 & a_4 \\ a_5 & a_6 & a_7 & u_8 \\ a_9 & a_{10} & a_{11} & a_{12} \\ a_{13} & a_{14} & a_{15} & a_{16} \end{array} \right] \begin{array}{c} 1 \\ 2 \\ 3 \\ 4 \end{array} \end{array} \left\{ \begin{array}{c} D_1^1 \\ D_2^1 \\ D_3^1 \\ D_4^1 \end{array} \right\}
\tag{a}
$$

Element 2

$$\{P^2\} = [K^2]\{D^2\}$$

or

$$
\left\{\begin{array}{c} P_3^2 \\ P_4^2 \\ P_5^2 \\ P_6^2 \end{array}\right\} = \begin{array}{c} \begin{array}{cccc} 3 & 4 & 5 & 6 \end{array} \\ \left[\begin{array}{cccc} b_1 & b_2 & b_3 & b_4 \\ b_5 & b_6 & b_7 & b_8 \\ b_9 & b_{10} & b_{11} & b_{12} \\ b_{13} & b_{14} & b_{15} & b_{16} \end{array}\right] \begin{array}{c} 3 \\ 4 \\ 5 \\ 6 \end{array} \end{array} \left\{\begin{array}{c} D_3^2 \\ D_4^2 \\ D_5^2 \\ D_6^2 \end{array}\right\} \tag{b}
$$

Element 3

$$
\{P^3\} = [K^3]\{D^3\}
$$

or

$$
\left\{\begin{array}{c} P_1^3 \\ P_2^3 \\ P_5^3 \\ P_6^3 \end{array}\right\} = \begin{array}{c} \begin{array}{cccc} 1 & 2 & 5 & 6 \end{array} \\ \left[\begin{array}{cccc} c_1 & c_2 & c_3 & c_4 \\ c_5 & c_6 & c_7 & c_8 \\ c_9 & c_{10} & c_{11} & c_{12} \\ c_{13} & c_{14} & c_{15} & c_{16} \end{array}\right] \begin{array}{c} 1 \\ 2 \\ 5 \\ 6 \end{array} \end{array} \left\{\begin{array}{c} D_1^3 \\ D_2^3 \\ D_5^3 \\ D_6^3 \end{array}\right\} \tag{c}
$$

The compatibility conditions at the nodes are

At Node 1:

$$
D_1 = D_1^1 = D_1^3
$$

$$
D_2 = D_2^1 = D_2^3
$$

At Node 2:

$$
D_3 = D_3^1 = D_3^2
$$

$$
D_4 = D_4^1 = D_4^2
$$

At Node 3:

$$
D_5 = D_5^2 = D_5^3
$$

$$
D_6 = D_6^2 = D_6^3
$$

$$
\tag{d}
$$

The FBD of the nodes are shown in Figure 5.7d. Using these FBD, the following equilibrium equations are obtained:

$$
P_1 = P_1^1 + P_1^3
$$

$$
P_2 = P_2^1 + P_2^3
$$

$$
P_3 = P_3^1 + P_3^2
$$

$$
P_4 = P_4^1 + P_4^2
$$

$$
P_5 = P_5^2 + P_5^3
$$

$$
P_6 = P_6^2 + P_6^3
$$

$$
\tag{e}
$$

Substituting Eqs. (a) to (d) in Eq. (e) gives

$$
\begin{Bmatrix} P_1 \\ P_2 \\ P_3 \\ P_4 \\ P_5 \\ P_6 \end{Bmatrix}
=
\begin{Bmatrix} P_1^1 + P_1^3 \\ P_2^1 + P_2^3 \\ P_3^1 + P_3^2 \\ P_4^1 + P_4^2 \\ P_5^2 + P_5^3 \\ P_6^2 + P_6^3 \end{Bmatrix}
$$

$$
=
\begin{bmatrix}
a_1 + c_1 & a_2 + c_2 & a_3 & a_4 & c_3 & c_4 \\
a_5 + c_5 & a_6 + c_6 & a_7 & a_8 & c_7 & c_8 \\
a_9 & a_{10} & a_{11} + b_1 & a_{12} + b_2 & b_3 & b_4 \\
a_{13} & a_{14} & a_{15} + b_5 & a_{16} + b_6 & b_7 & b_8 \\
c_9 & c_{10} & b_9 & b_{10} & b_{11} + c_{11} & b_{12} + c_{12} \\
c_{13} & c_{14} & b_{13} & b_{14} & b_{15} + c_{15} & b_{16} + c_{16}
\end{bmatrix}
\begin{Bmatrix} D_1 \\ D_2 \\ D_3 \\ D_4 \\ D_5 \\ D_6 \end{Bmatrix}
$$

(f)

or

$$\{P\} = [K]\{D\}$$

where $[K]$ is the structure stiffness matrix.

5.6.1 Procedure for Developing [K]

The above method for obtaining $[K]$, i.e., by using compatibility and equilibrium equations at the nodes, will be time consuming. The element stiffness matrices with respect to global axes can be easily assembled by identifying the global coordinates of the near node N and far node F of the element. Since the example considered is a truss element, each node will have two global coordinates: the global coordinates of the near node and far node along the X- and Y-axes are denoted as N_1, N_2, and F_1, F_2, respectively. The coordinate labels for the nodes are tabulated in Table 5.1.

TABLE 5.1

Coordinates of Near and Far Node

	Nodes		Global Coordinates			
Element	Near (N)	Far (F)	N_1	N_2	F_1	F_2
1	1	2	1	2	3	4
2	2	3	3	4	5	6
3	1	3	1	2	5	6

The rows and columns of the element stiffness matrices are numbered using the details of the global coordinates given in Table 5.1 (see the numbering of $[K^i]$ in Eqs. (a) to (c)). The row and column of $[K]$ to which the elements of $[K^i]$ are to be placed is identified using the coordinate labels written on the right and at top of $[K^i]$. For example, the stiffness coefficient a_{12} in $[K^i]$ (Eq. a) is to be placed in the third row and fourth column of $[K]$. While placing the stiffness coefficients from $[K^i]$ to $[K]$, if more than one stiffness coefficient from the element stiffness matrices occupies the same location in $[K]$, then the stiffness coefficients are to be added. The physical significance of adding the stiffness coefficients is that these coefficients indicate the resistance of element to applied loads. Thus, if a load is applied at a node of structure, then all the elements joining at the node should resist the load. Hence, the added stiffness coefficients indicate the total resistance of the node.

5.7 Procedure for Analyzing Framed Structure Using Direct Stiffness Method

The steps for analysis of structures using the direct stiffness method are outlined below:

i. Develop the discretized structure and label the nodes and elements.
ii. Identify the active and restrained coordinates. Define the load vector $\{P_A\}$, reaction vector $\{P_R\}$, active displacement vector, $\{D_A\}$ and restrained displacement vector $\{D_R\}$.
iii. For each element, identify the coordinates with respect to local and global axes. Also define the element displacement vectors in local and global coordinates ($\{d^i\}$ and $\{D^i\}$), and element force vectors in local and global coordinates ($\{p^i\}$ and $\{P^i\}$).
iv. Develop the element stiffness matrix with respect to local axes $[k^i]$ and element transformation matrix $[T]$.
v. Transform the element stiffness matrix with respect to local axes $[k^i]$ to global axes $[K^i]$ using the equation $[K^i] = [T]^T[k^i][T]$.
vi. Assemble $[K^i]$ of all elements to develop the structure stiffness matrix $[K]$. Partition $[K]$ into four submatrices of the form $[K]=\begin{bmatrix} [K_{AA}] & [K_{AR}] \\ [K_{RA}] & [K_{RR}] \end{bmatrix}$.
vii. Determine $\{D_A\}$ using the equation $\{D_A\} = [K_{AA}]^{-1}(\{P_A\} - [K_{AR}]\{D_R\})$.
viii. Find the reaction vector $\{P_R\}$ by equation $\{P_R\} = [K_{RA}]\{D_A\} + [K_{RR}]\{D_R\}$.
ix. Calculate the element force vector in local coordinates $\{p^i\}$ using Eq. 5.18.
$\{p^i\} = [k^i][T]\{D^i\}$

5.7.1 Analysis of Structures with Element Loads

For these types of structures, the combined force vectors along the active and restrained coordinates ($\{P_{CA}\}$ and $\{P_{CR}\}$) should be used instead of $\{P_A\}$ and $\{P_R\}$. The element force vector in local coordinates is calculated using the following equation:

$$\{p^i\} = \{p^i_F\} + [k^i][T^i]\{D^i\} \qquad (5.19)$$

where $\{p^i_F\}$ is the force developed in element i due to fixed end actions.

Example 5.1

Analyze the truss shown in Figure 5.9a using the direct stiffness method.

Solution

Step 1: Develop the discretized structure and identify global coordinates (active and restrained)

The elements and nodes of truss are shown in Figure 5.9b. Node 2 is chosen as the origin of the global axes. The active (1 and 2) and restrained coordinates (3 to 6) and the global axes of the structure are shown in Figure 5.9c.

Step 2: Identify the force and displacement vectors of the structure.

- Load vector $\{P_A\}$

$$\{P_A\} = \left\{ \begin{array}{c} P_1 \\ P_2 \end{array} \right\} = \left\{ \begin{array}{c} 0 \\ -20 \end{array} \right\} kN$$

- Reaction vector $\{P_R\}$

$$\{P_R\} = \left\{ \begin{array}{c} P_3 \\ P_4 \\ P_5 \\ P_6 \end{array} \right\}$$

- Active displacement vector $\{D_A\}$

$$\{D_A\} = \left\{ \begin{array}{c} D_1 \\ D_2 \end{array} \right\}$$

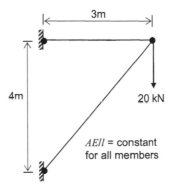

FIGURE 5.9a
Plane truss.

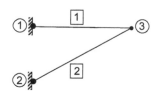

FIGURE 5.9b
Elements and nodes.

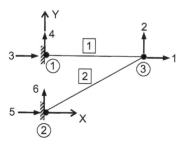

FIGURE 5.9c
Global axes and coordinates.

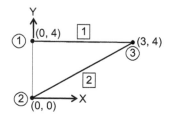

FIGURE 5.9d
Coordinate of node.

- Restrained displacement vector $\{D_R\}$

$$\{D_R\} = \left\{ \begin{array}{c} D_3 \\ D_4 \\ D_5 \\ D_6 \end{array} \right\} = \left\{ \begin{array}{c} 0 \\ 0 \\ 0 \\ 0 \end{array} \right\}$$

Step 3: Develop element stiffness matrix

- Coordinates of node and element connectivity data
  The coordinates of the nodes are shown in Figure 5.9d. The details are also given in Table 5.2. The element connectivity data are tabulated in Table 5.3.
- Local axes and coordinates
  Figure 5.9e shows the local axes and local coordinates of the elements.
- Element stiffness matrix for element 1
  Element stiffness matrix in local coordinates $[k^i]$

TABLE 5.2

Coordinates of Nodes

Node	X (m)	Y (m)
1	0	4
2	0	0
3	3	4

TABLE 5.3

Element Connectivity Data

Element	Near Node N	Far Node F
1	1	3
2	2	3

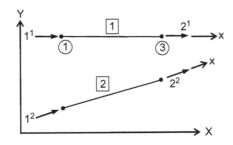

FIGURE 5.9e
Local axes and local coordinates.

$$[k^1] = \frac{AE}{l}\begin{bmatrix} 1 & -1 \\ -1 & 1 \end{bmatrix}$$

Element transformation matrix $[T^i]$

$$l = \sqrt{(X_F - X_N)^2 + (Y_F - Y_N)^2} = \sqrt{(3-0)^2 + (4-4)^2} = 3 \text{ m}$$

$$c = \frac{X_F - X_N}{l} = \frac{3-0}{3} = 1$$

$$s = \frac{Y_F - Y_N}{l} = \frac{(4-4)}{3} = 0$$

$$[T^1] = \begin{bmatrix} c & s & 0 & 0 \\ 0 & 0 & c & s \end{bmatrix} = \begin{bmatrix} 1 & 0 & 0 & 0 \\ 0 & 0 & 1 & 0 \end{bmatrix}$$

Element stiffness matrix in global coordinates $[K^i]$

$$\left[K^i\right]=\left[T^1\right]^T\left[k^1\right]\left[T^1\right]$$

$$=\frac{AE}{l}\begin{array}{cccc} 3 & 4 & 1 & 2 \\ \left[\begin{array}{cccc} 1 & 0 & -1 & 0 \\ 0 & 0 & 0 & 0 \\ -1 & 0 & 1 & 0 \\ 0 & 0 & 0 & 0 \end{array}\right] & & & \begin{array}{c} 3 \\ 4 \\ 1 \\ 2 \end{array}\end{array}$$

- Element stiffness matrix for element 2
  Element stiffness matrix in local coordinates $[k^i]$

$$\left[k^1\right]=\frac{AE}{l}\left[\begin{array}{cc} 1 & -1 \\ -1 & 1 \end{array}\right]$$

Element transformation matrix $[T]$

$$l=\sqrt{(3-0)^2+(4-0)^2}=5\text{ m}$$

$$c=\frac{X_F-X_N}{l}=\frac{3}{5}=0.6$$

$$s=\frac{Y_F-Y_N}{l}=\frac{4}{5}=0.8$$

$$\left[T^2\right]=\left[\begin{array}{cccc} 0.6 & 0.8 & 0 & 0 \\ 0 & 0 & 0.6 & 0.8 \end{array}\right]$$

Element stiffness matrix in global coordinates $[K^i]$

$$\left[K^2\right]=\frac{AE}{l}\begin{array}{cccc} 5 & 6 & 1 & 2 \\ \left[\begin{array}{cccc} 0.36 & 0.48 & -0.36 & -0.48 \\ 0.48 & 0.64 & -0.48 & -0.64 \\ -0.36 & -0.48 & 0.36 & 0.48 \\ -0.48 & -0.64 & 0.48 & 0.64 \end{array}\right] & & & \begin{array}{c} 5 \\ 6 \\ 1 \\ 2 \end{array}\end{array}$$

Step 4: Develop structure stiffness matrix $[K]$

$$[K]=\frac{AE}{l}\begin{array}{cccccc} 1 & 2 & 3 & 4 & 5 & 6 \\ \left[\begin{array}{cccccc} 1+0.36 & 0+0.48 & -1 & 0 & -0.36 & -0.48 \\ 0+0.48 & 0+0.64 & 0 & 0 & -0.48 & -0.64 \\ -1 & 0 & 1 & 0 & 0 & 0 \\ 0 & 0 & 0 & 0 & 0 & 0 \\ -0.36 & -0.48 & 0 & 0 & 0.36 & 0.48 \\ -0.48 & -0.64 & 0 & 0 & 0.48 & 0.64 \end{array}\right] & & & & & \begin{array}{c} 1 \\ 2 \\ 3 \\ 4 \\ 5 \\ 6 \end{array}\end{array}$$

$$\text{Thus, } [K] = \left[\begin{array}{c|c} [K_{AA}] & [K_{AR}] \\ \hline [K_{RA}] & [K_{RR}] \end{array} \right] = \frac{AE}{l} \left[\begin{array}{cc|cccc} 1.36 & 0.48 & -1 & 0 & -0.36 & -0.48 \\ 0.48 & 0.64 & 0 & 0 & -0.48 & -0.64 \\ \hline -1 & 0 & 1 & 0 & 0 & 0 \\ 0 & 0 & 0 & 0 & 0 & 0 \\ -0.36 & -0.48 & 0 & 0 & 0.36 & 0.48 \\ -0.48 & -0.64 & 0 & 0 & 0.48 & 0.64 \end{array} \right]$$

Step 5: Find $\{D_A\}$

$$\{D_A\} = [K_{AA}]^{-1}\left(\{P_A\} - [K_{AR}]\{D_R\}\right)$$

Since $\{D_R\} = \{0\}$,

$$\{D_A\} = [K_{AA}]^{-1}\{P_A\}$$

$$[K_{AA}] = \frac{AE}{l}\left[\begin{array}{cc} 1.36 & 0.48 \\ 0.48 & 0.64 \end{array} \right]$$

$$\text{Therefore, } \{D_A\} = \frac{l}{AE}\left[\begin{array}{c} 15 \\ -42.5 \end{array} \right]$$

Step 6: Find $\{P_R\}$

$$\{D_R\} = \{0\}$$

$$\{P_R\} = [K_{RA}]\{D_A\}$$

$$[K_{RA}] = \frac{AE}{l}\left[\begin{array}{cc} -1 & 0 \\ 0 & 0 \\ -0.36 & -0.48 \\ -0.48 & -0.64 \end{array} \right]$$

$$\text{Therefore, } \{P_R\} = \left\{ \begin{array}{c} -15 \\ 0 \\ 15 \\ 20 \end{array} \right\} kN$$

Step 7: Determine the element force vector in local coordinates.

- Element displacement vector in global coordinates
  The displacements of elements in global coordinates are shown in Figure 5.9f.

$$\{D^1\} = \left\{ \begin{array}{c} D_3^1 \\ D_4^1 \\ D_1^1 \\ D_2^1 \end{array} \right\} = \left\{ \begin{array}{c} D_3 \\ D_4 \\ D_1 \\ D_2 \end{array} \right\} = \frac{l}{AE}\left\{ \begin{array}{c} 0 \\ 0 \\ 15 \\ -42.5 \end{array} \right\}$$

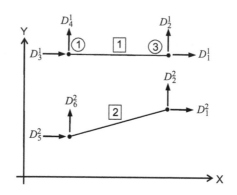

FIGURE 5.9f
Displacements of elements in global coordinates.

$$\{D^2\} = \begin{Bmatrix} D_5^2 \\ D_6^2 \\ D_1^2 \\ D_2^2 \end{Bmatrix} = \begin{Bmatrix} D_5 \\ D_6 \\ D_1 \\ D_2 \end{Bmatrix} = \frac{l}{AE} \begin{Bmatrix} 0 \\ 0 \\ 15 \\ -42.5 \end{Bmatrix}$$

- Element force vector in local coordinates $\{p^i\}$

$$\{p^i\} = [k^i][T^i]\{D^i\}$$

Element 1

$$\{p^1\} = [k^1][T^1]\{D^1\} = \begin{Bmatrix} -15 \\ 15 \end{Bmatrix} kN$$

Element 2

$$\{p^2\} = [k^2][T^2]\{D^2\} = \begin{Bmatrix} 25 \\ -25 \end{Bmatrix} kN$$

The force in elements is shown in Figure 5.9g. The reactions are indicated in Figure 5.9h.

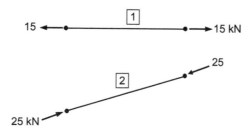

FIGURE 5.9g
Force in elements.

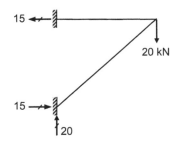

FIGURE 5.9h
Support reactions.

Example 5.2

Analyze the space truss shown in Figure 5.10a using the direct stiffness method. AE/l for all the elements is same.

Solution

Step 1: Develop the discretized structure and identify the global coordinates

The elements, nodes, nodal coordinates, and global axes are shown in Figure 5.10b. The active coordinates (1, 2, 3) and restrained coordinates (4 to 12) are shown in Figure 5.10c.

Step 2: Identify the force and displacement vectors of the structure

$$\{P_A\} = \begin{Bmatrix} P_1 \\ P_2 \\ P_3 \end{Bmatrix} = \begin{Bmatrix} 10 \\ -15 \\ 5 \end{Bmatrix} kN, \{D_A\} = \begin{Bmatrix} D_1 \\ D_2 \\ D_3 \end{Bmatrix}$$

$$\{P_R\} = \begin{Bmatrix} P_4 \\ P_5 \\ P_6 \\ P_7 \\ P_8 \\ P_9 \\ P_{10} \\ P_{11} \\ P_{12} \end{Bmatrix}, \{D_A\} = \begin{Bmatrix} D_4 \\ D_5 \\ D_6 \\ D_7 \\ D_8 \\ D_9 \\ D_{10} \\ D_{11} \\ D_{12} \end{Bmatrix} = \begin{Bmatrix} 0 \\ 0 \\ 0 \\ 0 \\ 0 \\ 0 \\ 0 \\ 0 \\ 0 \end{Bmatrix}$$

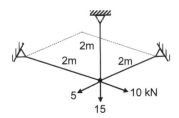

FIGURE 5.10a
Space truss.

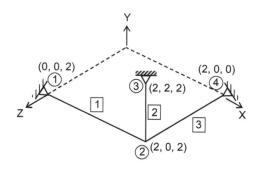

FIGURE 5.10b
Discretized structure.

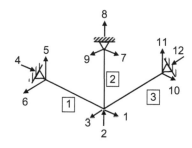

FIGURE 5.10c
Global coordinates.

Step 3: Develop element stiffness matrix

The nodal coordinates and element connectivity data are given in Tables 5.4 and 5.5. The local axes and coordinates of the elements are shown in Figure 5.10d.

- Element stiffness matrix in local coordinates

$$\left[k^1\right]=\left[k^2\right]=\left[k^3\right]=\frac{AE}{l}\begin{bmatrix} 1 & -1 \\ -1 & 1 \end{bmatrix}$$

- Element transformation matrix $[T^i]$

Element 1

$$l=\sqrt{(X_F-X_N)^2+(Y_F-Y_N)^2+(Z_F-Z_N)^2}=\sqrt{2^2+0^2+0^2}=2\text{ m}$$

$$l_1=\frac{X_F-X_N}{l}=1$$

$$m_1=\frac{Y_F-Y_N}{l}=0$$

$$n_1=\frac{Z_F-Z_N}{l}=0$$

$$\left[T^1\right]=\begin{bmatrix} l_1 & m_1 & n_1 & 0 & 0 & 0 \\ 0 & 0 & 0 & l_1 & m_1 & n_1 \end{bmatrix}=\begin{bmatrix} 1 & 0 & 0 & 0 & 0 & 0 \\ 0 & 0 & 0 & 1 & 0 & 0 \end{bmatrix}$$

TABLE 5.4

Coordinates of Node

Node	X (m)	Y (m)	Z (m)
1	0	0	2
2	2	0	2
3	2	2	2
4	2	0	0

TABLE 5.5

Element Connectivity Data

Element	Near Node N	Far Node F
1	1	2
2	2	3
3	2	4

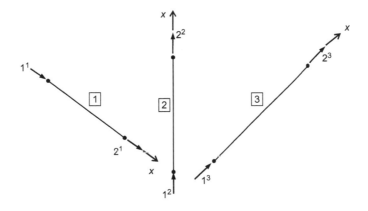

FIGURE 5.10d
Local axes and coordinates.

Element 2

$$l = \sqrt{0^2 + 2^2 + 0^2} = 2 \text{ m}$$

$$l_1 = 0,\ m_1 = 1,\ n_1 = 0$$

$$[T^2] = \begin{bmatrix} 0 & 1 & 0 & 0 & 0 & 0 \\ 0 & 0 & 0 & 0 & 1 & 0 \end{bmatrix}$$

Element 3

$$l = \sqrt{0^2 + 0^2 + (-2)^2} = 2 \text{ m}$$

$$l_1 = 0,\ m_1 = 0,\ n_1 = -1$$

$$[T^3] = \begin{bmatrix} 0 & 0 & -1 & 0 & 0 & 0 \\ 0 & 0 & 0 & 0 & 0 & -1 \end{bmatrix}$$

- Element stiffness matrix in global coordinates $[K^i]$

$$[K^i] = [T^i]^T [k^i][T^i]$$

$$[K^1] = \frac{AE}{l}
\begin{array}{c}
\begin{array}{cccccc} 4 & 5 & 6 & 1 & 2 & 3 \end{array} \\
\left[\begin{array}{cccccc}
1 & 0 & 0 & -1 & 0 & 0 \\
0 & 0 & 0 & 0 & 0 & 0 \\
0 & 0 & 0 & 0 & 0 & 0 \\
-1 & 0 & 0 & 1 & 0 & 0 \\
0 & 0 & 0 & 0 & 0 & 0 \\
0 & 0 & 0 & 0 & 0 & 0
\end{array}\right]
\begin{array}{c} 4 \\ 5 \\ 6 \\ 1 \\ 2 \\ 3 \end{array}
\end{array}$$

$$[K^2] = \frac{AE}{l}
\begin{array}{c}
\begin{array}{cccccc} 1 & 2 & 3 & 7 & 8 & 9 \end{array} \\
\left[\begin{array}{cccccc}
0 & 0 & 0 & 0 & 0 & 0 \\
0 & 1 & 0 & 0 & -1 & 0 \\
0 & 0 & 0 & 0 & 0 & 0 \\
0 & 0 & 0 & 0 & 0 & 0 \\
0 & -1 & 0 & 0 & 1 & 0 \\
0 & 0 & 0 & 0 & 0 & 0
\end{array}\right]
\begin{array}{c} 1 \\ 2 \\ 3 \\ 7 \\ 8 \\ 9 \end{array}
\end{array}$$

$$[K^3] = \frac{AE}{l}
\begin{array}{c}
\begin{array}{cccccc} 1 & 2 & 3 & 10 & 11 & 12 \end{array} \\
\left[\begin{array}{cccccc}
0 & 0 & 0 & 0 & 0 & 0 \\
0 & 0 & 0 & 0 & 0 & 0 \\
0 & 0 & 1 & 0 & 0 & -1 \\
0 & 0 & 0 & 0 & 0 & 0 \\
0 & 0 & 0 & 0 & 0 & 0 \\
0 & 0 & -1 & 0 & 0 & 1
\end{array}\right]
\begin{array}{c} 1 \\ 2 \\ 3 \\ 10 \\ 11 \\ 12 \end{array}
\end{array}$$

Step 4: Develop structure stiffness matrix

$$[K] = \frac{AE}{l}
\begin{array}{c}
\begin{array}{cccccccccccc} 1 & 2 & 3 & 4 & 5 & 6 & 7 & 8 & 9 & 10 & 11 & 12 \end{array} \\
\left[\begin{array}{ccc|ccccccccc}
1 & 0 & 0 & -1 & 0 & 0 & 0 & 0 & 0 & 0 & 0 & 0 \\
0 & 1 & 0 & 0 & 0 & 0 & 0 & -1 & 0 & 0 & 0 & 0 \\
0 & 0 & 1 & 0 & 0 & 0 & 0 & 0 & 0 & 0 & 0 & -1 \\ \hline
-1 & 0 & 0 & 1 & 0 & 0 & 0 & 0 & 0 & 0 & 0 & 0 \\
0 & 0 & 0 & 0 & 0 & 0 & 0 & 0 & 0 & 0 & 0 & 0 \\
0 & 0 & 0 & 0 & 0 & 0 & 0 & 0 & 0 & 0 & 0 & 0 \\
0 & 0 & 0 & 0 & 0 & 0 & 0 & 0 & 0 & 0 & 0 & 0 \\
0 & -1 & 0 & 0 & 0 & 0 & 0 & 1 & 0 & 0 & 0 & 0 \\
0 & 0 & 0 & 0 & 0 & 0 & 0 & 0 & 0 & 0 & 0 & 0 \\
0 & 0 & 0 & 0 & 0 & 0 & 0 & 0 & 0 & 0 & 0 & 0 \\
0 & 0 & 0 & 0 & 0 & 0 & 0 & 0 & 0 & 0 & 0 & 0 \\
0 & 0 & -1 & 0 & 0 & 0 & 0 & 0 & 0 & 0 & 0 & 1
\end{array}\right]
\begin{array}{c} 1 \\ 2 \\ 3 \\ 4 \\ 5 \\ 6 \\ 7 \\ 8 \\ 9 \\ 10 \\ 11 \\ 12 \end{array}
\end{array}$$

Step 5: Find $\{D_A\}$

$$\{D_A\} = \{0\}$$

$$\{D_A\} = [K_{AA}]^{-1}\{P_A\} = \frac{l}{AE}\begin{Bmatrix} 10 \\ -15 \\ 5 \end{Bmatrix}$$

Step 6: Find $\{P_R\}$

$$\{P_R\} = [K_{RA}]\{D_A\} = \begin{Bmatrix} -10 \\ 0 \\ 0 \\ 0 \\ 15 \\ 0 \\ 0 \\ 0 \\ -5 \end{Bmatrix} kN$$

Step 7: Determine element forces

$$\{D^1\} = \begin{Bmatrix} D_4 \\ D_5 \\ D_6 \\ D_1 \\ D_2 \\ D_3 \end{Bmatrix} = \frac{l}{AE}\begin{Bmatrix} 0 \\ 0 \\ 0 \\ 10 \\ -15 \\ 5 \end{Bmatrix}, \{D^2\} = \begin{Bmatrix} D_1 \\ D_2 \\ D_3 \\ D_7 \\ D_8 \\ D_9 \end{Bmatrix} = \frac{l}{AE}\begin{Bmatrix} 10 \\ -15 \\ 5 \\ 0 \\ 0 \\ 0 \end{Bmatrix},$$

$$\{D^3\} = \begin{Bmatrix} D_1 \\ D_2 \\ D_3 \\ D_{10} \\ D_{11} \\ D_{12} \end{Bmatrix} = \frac{l}{AE}\begin{Bmatrix} 10 \\ -15 \\ 5 \\ 0 \\ 0 \\ 0 \end{Bmatrix}$$

$$\{p^1\} = [k^1][T^1]\{D^1\} = \begin{Bmatrix} -10 \\ 10 \end{Bmatrix} kN$$

$$\{p^2\} = [k^2][T^2]\{D^2\} = \begin{Bmatrix} -15 \\ 15 \end{Bmatrix} kN$$

$$\{p^3\} = [k^3][T^3]\{D^3\} = \begin{Bmatrix} -5 \\ 5 \end{Bmatrix} kN$$

The FBD of the elements are shown in Figure 5.10e.

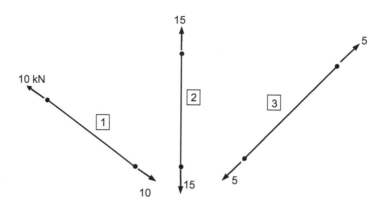

FIGURE 5.10E
FBD of elements.

Example 5.3

Determine the slope at point B of the beam shown in Figure 5.11a. Also find support reactions and forces developed in the beam. Use the direct stiffness method.

Solution

Step 1: Develop the discretized structure and identify global coordinates
   The elements, nodes, and global axes are indicated in Figure 5.11b. The active coordinates (1 to 4) and restrained coordinates (5 to 6) are shown in Figure 5.11c.
Step 2: Define the force and displacement vectors of the structure
   The fixed end actions and equivalent nodal loads are shown in Figures 5.11d and e, respectively.

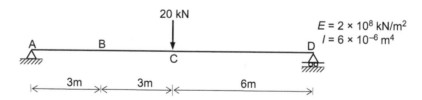

FIGURE 5.11a
Simply supported beam.

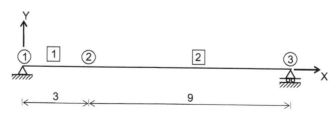

FIGURE 5.11b
Discretized structure.

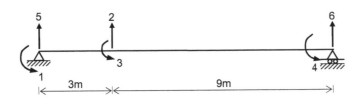

FIGURE 5.11c
Global coordinates.

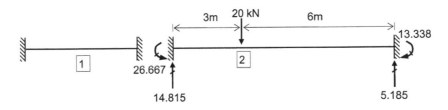

FIGURE 5.11d
Fixed end actions.

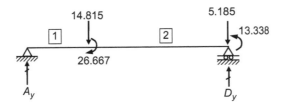

FIGURE 5.11e
Equivalent nodal loads.

The combined force vectors along active and restrained coordinates are

$$\{P_{CA}\} = \begin{Bmatrix} P_{c1} \\ P_{c2} \\ P_{c3} \\ P_{c4} \end{Bmatrix} = \begin{Bmatrix} 0 \\ -14.815 \\ -26.667 \\ 13.333 \end{Bmatrix}$$

$$\{P_{CR}\} = \begin{Bmatrix} P_{c5} \\ P_{c6} \end{Bmatrix} = \begin{Bmatrix} A_y \\ D_y - 5.185 \end{Bmatrix}$$

The corresponding displacement vectors are

$$\{D_A\} = \begin{Bmatrix} D_1 \\ D_2 \\ D_3 \\ D_4 \end{Bmatrix}; \{D_R\} = \begin{Bmatrix} D_5 \\ D_6 \end{Bmatrix} = \begin{Bmatrix} 0 \\ 0 \end{Bmatrix}$$

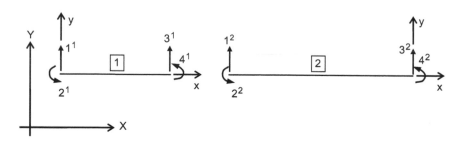

FIGURE 5.11f
Local axes and coordinates.

Step 3: Develop element stiffness matrix

In the case of beam elements, $[T]$ is an identity matrix and $[k^i] = [K^i]$. The local coordinates of the elements are shown in Figure 5.11f.

- Element 1; $l = 3\,\text{m}$

$$[k^1] = [K^1] = EI \begin{array}{c} \\ \\ \\ \\ \end{array} \begin{array}{cccc} 5 & 1 & 2 & 3 \\ \left[\begin{array}{cccc} 0.444 & 0.667 & -0.444 & 0.667 \\ 0.667 & 1.333 & -0.667 & 0.667 \\ -0.444 & -0.667 & 0.444 & -0.667 \\ 0.667 & 0.667 & -0.667 & 1.333 \end{array} \right] & \begin{array}{c} 5 \\ 1 \\ 2 \\ 3 \end{array} \end{array}$$

- Element 2; $l = 9\,\text{m}$

$$[k^2] = [K^2] = EI \begin{array}{c} \\ \\ \\ \\ \end{array} \begin{array}{cccc} 2 & 3 & 6 & 4 \\ \left[\begin{array}{cccc} 0.0165 & 0.0741 & -0.0165 & 0.0741 \\ 0.0741 & 0.4444 & -0.0741 & 0.2222 \\ -0.0165 & -0.0741 & 0.0165 & -0.0741 \\ 0.0741 & 0.2222 & -0.0741 & 0.4444 \end{array} \right] & \begin{array}{c} 2 \\ 3 \\ 6 \\ 4 \end{array} \end{array}$$

Step 4: Develop structure stiffness matrix $[K]$

$$[K] = EI \begin{array}{c} \\ \\ \\ \\ \\ \\ \end{array} \begin{array}{cccccc} 1 & 2 & 3 & 4 & 5 & 6 \\ \left[\begin{array}{cccc|cc} 1.333 & -0.667 & 0.667 & 0 & 0.667 & 0 \\ -0.667 & 0.461 & -0.593 & 0.0741 & -0.444 & -0.0165 \\ 0.667 & -0.593 & 1.778 & 0.222 & 0.667 & -0.0741 \\ \hline 0 & 0.0741 & 0.222 & 0.444 & 0 & -0.0741 \\ 0.667 & -0.444 & 0.667 & 0 & 0.444 & 0 \\ 0 & -0.0165 & -0.0741 & -0.0741 & 0 & 0.0165 \end{array} \right] & \begin{array}{c} 1 \\ 2 \\ 3 \\ 4 \\ 5 \\ 6 \end{array} \end{array}$$

Step 5: Find $\{D_A\}$

$$\{D_R\} = \{0\}$$

$$\{D_A\} = [K_{AA}]^{-1}\{P_{CA}\} = \frac{1}{EI} \left\{ \begin{array}{c} -181.5304 \\ -498.5981 \\ -135.8095 \\ 181.1460 \end{array} \right\}$$

Rotation at $B = D_3 = \dfrac{-135.8095}{EI} = -0.113$ rad

Step 6: Find $\{P_{CR}\}$

$$\{P_{CR}\} = [K_{RA}]\{D_A\} = \left\{ \begin{array}{c} P_{c5} \\ P_{c6} \end{array} \right\} = \left\{ \begin{array}{c} 9.7118 \\ 4.8674 \end{array} \right\} \text{kN}$$

$$P_{c5} = A_y = 9.7118 \text{ kN}$$

$$P_{c6} = B_y - 5.185 = 4.8674 \Rightarrow B_y = 4.874 + 5.185 = 10.0524 \text{ kN}$$

Note: The correct values of A_y and B_y are 10 kN. The exact values can be obtained by taking more significant digits in calculations.

Step 7: Determine element forces

The element forces are determined using Eq. 5.19.

$$\{p^i\} = \{p^i_F\} + [k^i][T^i]\{D^i\}.$$

Element forces due to fixed end actions $\{p^i_F\}$

Comparing Figure 5.11d with Figure 5.11f gives

$$\{p_F^1\} = \left\{ \begin{array}{c} 0 \\ 0 \\ 0 \\ 0 \end{array} \right\}; \{p_F^2\} = \left\{ \begin{array}{c} 14.815 \\ 26.667 \\ 5.185 \\ -13.333 \end{array} \right\}$$

- Displacement vectors of elements in global coordinates $\{D^i\}$

$$\{D^1\} = \left\{ \begin{array}{c} D_5 \\ D_1 \\ D_2 \\ D_3 \end{array} \right\} = \dfrac{l}{EI} \left\{ \begin{array}{c} 0 \\ -181.5304 \\ -498.5981 \\ -135.8095 \end{array} \right\},$$

$$\{D^2\} = \left\{ \begin{array}{c} D_2 \\ D_3 \\ D_6 \\ D_4 \end{array} \right\} = \dfrac{l}{EI} \left\{ \begin{array}{c} 498.5981 \\ -135.8095 \\ 0 \\ 181.1460 \end{array} \right\}$$

- Element forces (in kN, m)

Element 1

$$\{p^1\} = \left\{ \begin{array}{c} 9.7118 \\ 0 \\ -9.7118 \\ 30.4501 \end{array} \right\}$$

Element 2

$$\{p^2\} = \begin{Bmatrix} 9.9476 \\ -30.3822 \\ 10.0524 \\ 0.0453 \end{Bmatrix}$$

Example 5.4

Analyze the plane frame shown in Figure 5.12a using the direct stiffness method.

Solution

Step 1: Develop the discretized structure and identify the global coordinates

The elements, nodes, and the nodal coordinates are indicated in Figure 5.12b. The active coordinates (1 to 6) and restrained coordinates (7 to 9) are shown in Figure 5.12c.

Step 2: Identify the force and displacement vectors of the structure

$$\{P_A\} = \begin{Bmatrix} P_1 \\ P_2 \\ P_3 \\ P_4 \\ P_5 \\ P_6 \end{Bmatrix} = \begin{Bmatrix} 0 \\ 0 \\ 0 \\ 0 \\ -10\,\text{kN} \\ 0 \end{Bmatrix} ; \{D_A\} = \begin{Bmatrix} D_1 \\ D_2 \\ D_3 \\ D_4 \\ D_5 \\ D_6 \end{Bmatrix}$$

$$\{P_R\} = \begin{Bmatrix} P_7 \\ P_8 \\ P_9 \end{Bmatrix} ; \{D_R\} = \begin{Bmatrix} D_7 \\ D_8 \\ D_9 \end{Bmatrix} = \begin{Bmatrix} 0 \\ 0 \\ 0 \end{Bmatrix}$$

Step 3: Develop element stiffness matrix

The coordinates of the nodes and the element connectivity data are tabulated in Tables 5.6 and 5.7. The local axes and to coordinates of the elements are shown in Figure 5.12d.

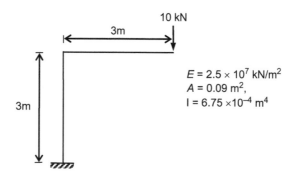

FIGURE 5.12a
Plane frame.

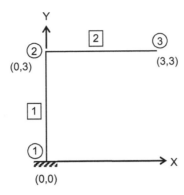

FIGURE 5.12b
Discretized structure.

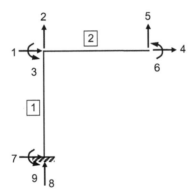

FIGURE 5.12c
Global coordinates.

TABLE 5.6

Coordinates of Node

Node	X (m)	Y (m)
1	0	0
2	0	3
3	3	3

TABLE 5.7

Element Connectivity Data

Element	Near Node N	Far Node F
1	1	2
2	2	3

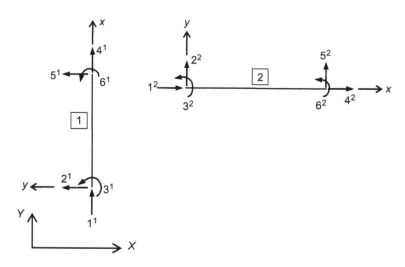

FIGURE 5.12d
Local axes and coordinates.

- Element stiffness matrix in local coordinates $[k^i]$
  Using Eq. 5.15c,

$$[k^1]=[k^2]=\begin{bmatrix} 750000 & 0 & 0 & -750000 & 0 & 0 \\ 0 & 7500 & 11250 & 0 & -7500 & 11250 \\ 0 & 11250 & 22500 & 0 & -11250 & 11250 \\ -750000 & 0 & 0 & 750000 & 0 & 0 \\ 0 & -7500 & -11250 & 0 & 7500 & -11250 \\ 0 & 11250 & 11250 & 0 & -11250 & 22500 \end{bmatrix}$$

- Element transformation matrix $[T^i]$

Element 1

$$l=\sqrt{(X_F-X_N)^2+(Y_F-Y_N)^2}=\sqrt{0^2+3^2}=3$$

$$c=\frac{X_F-X_N}{l}=0; s=\frac{Y_F-Y_N}{l}=\frac{3}{3}=1$$

Using Eq. 5.15b,

$$[T^1]=\begin{bmatrix} 0 & 1 & 0 & 0 & 0 & 0 \\ -1 & 0 & 0 & 0 & 0 & 0 \\ 0 & 0 & 1 & 0 & 0 & 0 \\ 0 & 0 & 0 & 0 & 1 & 0 \\ 0 & 0 & 0 & -1 & 0 & 0 \\ 0 & 0 & 0 & 0 & 0 & 1 \end{bmatrix}$$

Element 2

$$l=\sqrt{3^2+0^2}=1; c=1; s=0$$

$$[T^2] = \begin{bmatrix} 1 & 0 & 0 & 0 & 0 & 0 \\ 0 & 1 & 0 & 0 & 0 & 0 \\ 0 & 0 & 1 & 0 & 0 & 0 \\ 0 & 0 & 0 & 1 & 0 & 0 \\ 0 & 0 & 0 & 0 & 1 & 0 \\ 0 & 0 & 0 & 0 & 0 & 1 \end{bmatrix}$$

- Element stiffness matrix in global coordinates $[K^i]$

$$[K^1] = \begin{array}{c} \begin{array}{cccccc} 7 & 8 & 9 & 1 & 2 & 3 \end{array} \\ \begin{bmatrix} 7500 & 0 & -11250 & -7500 & 0 & -11250 \\ 0 & 750000 & 0 & 0 & -750000 & 0 \\ -11250 & 0 & 22500 & 11250 & 0 & 11250 \\ -7500 & 0 & 11250 & 7500 & 0 & 11250 \\ 0 & -750000 & 0 & 0 & 750000 & 0 \\ -11250 & 0 & 11250 & 11250 & 0 & 22500 \end{bmatrix} \begin{array}{c} 7 \\ 8 \\ 9 \\ 1 \\ 2 \\ 3 \end{array} \end{array}$$

$$[K^2] = \begin{array}{c} \begin{array}{cccccc} 1 & 2 & 3 & 4 & 5 & 6 \end{array} \\ \begin{bmatrix} 750000 & 0 & 0 & -750000 & 0 & 0 \\ 0 & 7500 & 11250 & 0 & -7500 & 11250 \\ 0 & 11250 & 22500 & 0 & -11250 & 11250 \\ -750000 & 0 & 0 & 750000 & 0 & 0 \\ 0 & -7500 & -11250 & 0 & 7500 & -11250 \\ 0 & 11250 & 11250 & 0 & -11250 & 22500 \end{bmatrix} \begin{array}{c} 1 \\ 2 \\ 3 \\ 4 \\ 5 \\ 6 \end{array} \end{array}$$

Step 4: Develop structure stiffness matrix $[K]$

$$[K] = \begin{array}{c} \begin{array}{ccccccccc} 1 & 2 & 3 & 4 & 5 & 6 & 7 & 8 & 9 \end{array} \\ \begin{bmatrix} 757500 & 0 & 11250 & -750000 & 0 & 0 & -7500 & 0 & 11250 \\ 0 & 757500 & 11250 & 0 & -7500 & 11250 & 0 & -750000 & 0 \\ 11250 & 11250 & 45000 & 0 & -11250 & 11250 & -11250 & 0 & 11250 \\ -750000 & 0 & 0 & 750000 & 0 & 0 & 0 & 0 & 0 \\ 0 & -7500 & -11250 & 0 & 7500 & -11250 & 0 & 0 & 0 \\ 0 & 11250 & 11250 & 0 & -11250 & 22500 & 0 & 0 & 0 \\ -7500 & 0 & -11250 & 0 & 0 & 0 & 7500 & 0 & -11250 \\ 0 & -750000 & 0 & 0 & 0 & 0 & 0 & 750000 & 0 \\ 11250 & 0 & 11250 & 0 & 0 & 0 & -11250 & 0 & 22500 \end{bmatrix} \begin{array}{c} 1 \\ 2 \\ 3 \\ 4 \\ 5 \\ 6 \\ 7 \\ 8 \\ 9 \end{array} \end{array}$$

Step 5: Find $\{D_A\}$

$$\{D_R\} = \{0\}$$

$$\{D_A\} = [K_{AA}]^{-1}\{P_A\} = \begin{Bmatrix} D_1 \\ D_2 \\ D_3 \\ D_4 \\ D_5 \\ D_6 \end{Bmatrix} = \begin{Bmatrix} 0.008 \\ -0.0000133 \\ -0.0053333 \\ 0.008 \\ -0.0213467 \\ -0.008 \end{Bmatrix}$$

Step 6: Find $\{P_R\}$

$$\{P_R\} = [K_{RA}]\{D_A\} = \begin{Bmatrix} 0 \\ 10 \\ 30 \end{Bmatrix}$$

Step 7: Determine element forces

$$\{D^1\} = \begin{Bmatrix} D_7 \\ D_8 \\ D_9 \\ D_1 \\ D_2 \\ D_3 \end{Bmatrix} = \begin{Bmatrix} 0 \\ 0 \\ 0 \\ 0.008 \\ -0.0000133 \\ -0.0053333 \end{Bmatrix} ; \{D^2\} = \begin{Bmatrix} D_1 \\ D_2 \\ D_3 \\ D_4 \\ D_5 \\ D_6 \end{Bmatrix} = \begin{Bmatrix} 0.008 \\ -0.0000133 \\ -0.0053333 \\ 0.008 \\ -0.0213467 \\ -0.008 \end{Bmatrix}$$

$$\{p^1\} = [k^1][T^1]\{D^1\} = \begin{Bmatrix} 10 \\ 0 \\ 30 \\ -10 \\ 0 \\ -30 \end{Bmatrix}$$

$$\{p^2\} = [k^2][T^2]\{D^2\} = \begin{Bmatrix} 0 \\ 10 \\ 30 \\ 0 \\ -10 \\ 0 \end{Bmatrix}$$

The FBD of the elements are shown in Figure 5.12e.

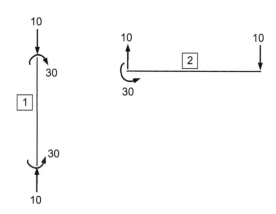

FIGURE 5.12e
FBD of elements (force in kN, m).

Example 5.5

Analyze the grid shown in Figure 5.13a using the direct stiffness method.

Solution

Step 1: Develop the discretized structure and identify the global coordinates

The discretized structure and global coordinates are indicated in Figures 5.13b and c. The global coordinates 1 to 3 are active coordinates and remaining are restrained coordinates.

Step 2: Identify the force and displacement vectors of the structure

$$\{P_A\} = \begin{Bmatrix} P_1 \\ P_2 \\ P_3 \end{Bmatrix} = \begin{Bmatrix} -10 \text{ kN} \\ 0 \\ 0 \end{Bmatrix}; \{D_A\} = \begin{Bmatrix} D_1 \\ D_2 \\ D_3 \end{Bmatrix}$$

$$\{P_R\} = \begin{Bmatrix} P_4 \\ P_5 \\ P_6 \\ P_7 \\ P_8 \\ P_9 \end{Bmatrix}; \{D_R\} = \begin{Bmatrix} D_4 \\ D_5 \\ D_6 \\ D_7 \\ D_8 \\ D_9 \end{Bmatrix} = \begin{Bmatrix} 0 \\ 0 \\ 0 \\ 0 \\ 0 \\ 0 \end{Bmatrix}$$

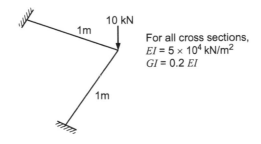

For all cross sections,
$EI = 5 \times 10^4 \text{ kN/m}^2$
$GI = 0.2\ EI$

FIGURE 5.13a
Grid.

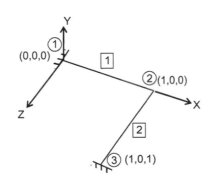

FIGURE 5.13b
Discretized structure.

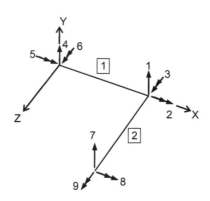

FIGURE 5.13c
Global coordinates.

Step 3: Develop element stiffness matrix

The nodal coordinates and element connectivity data are given in Tables 5.8 and 5.9. The local coordinates of the elements are shown in Figure 5.13d.

- Element stiffness matrix in local coordinates [k^i]

  Using Eq. 5.16c,

$$[k^1] = [k^2] = EI \begin{bmatrix} 12 & 0 & 6 & -12 & 0 & 6 \\ 0 & 0.2 & 0 & 0 & -0.2 & 0 \\ 6 & 0 & 4 & -6 & 0 & 2 \\ -12 & 0 & -6 & 12 & 0 & -6 \\ 0 & -0.2 & 0 & 0 & 0.2 & 0 \\ 6 & 0 & 2 & -6 & 0 & 4 \end{bmatrix}$$

- Element transformation matrix [T^i]

TABLE 5.8

Coordinates of Node

Node	X (m)	Y (m)	Z (m)
1	0	0	0
2	1	0	0
3	1	0	1

TABLE 5.9

Element Connectivity Data

Element	Near Node N	Far Node F
1	1	2
2	2	3

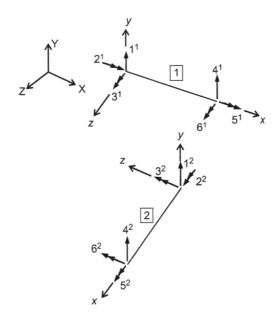

FIGURE 5.13d
Local axes and coordinates.

Element 1

$$l = \sqrt{(X_F - X_N)^2 + (Y_F - Y_N)^2} = \sqrt{1^2 + 0^2} = 1$$

$$\cos\alpha = \frac{X_F - X_N}{l} = 1; \sin\alpha = \frac{-(Z_F - Z_N)}{l} = 0$$

$$[T^1] = \begin{bmatrix} 1 & 0 & 0 & 0 & 0 & 0 \\ 0 & 1 & 0 & 0 & 0 & 0 \\ 0 & 0 & 1 & 0 & 0 & 0 \\ 0 & 0 & 0 & 1 & 0 & 0 \\ 0 & 0 & 0 & 0 & 1 & 0 \\ 0 & 0 & 0 & 0 & 0 & 1 \end{bmatrix}$$

Element 2

$$l = \sqrt{0^2 + 1^2} = 1; \cos\alpha = 0; \sin\alpha = -1$$

$$[T^2] = \begin{bmatrix} 1 & 0 & 0 & 0 & 0 & 0 \\ 0 & 0 & 1 & 0 & 0 & 0 \\ 0 & -1 & 0 & 0 & 0 & 0 \\ 0 & 0 & 0 & 1 & 0 & 0 \\ 0 & 0 & 0 & 0 & 0 & 1 \\ 0 & 0 & 0 & 0 & -1 & 0 \end{bmatrix}$$

- Element stiffness matrix in global coordinates $[K^i]$

$$[K^1] = EI \begin{array}{c} \\ \end{array} \begin{array}{cccccc} 4 & 5 & 6 & 1 & 2 & 3 \\ \left[\begin{array}{cccccc} 12 & 0 & 6 & -12 & 0 & 6 \\ 0 & 0.2 & 0 & 0 & -0.2 & 0 \\ 6 & 0 & 4 & -6 & 0 & 2 \\ -12 & 0 & -6 & 12 & 0 & -6 \\ 0 & -0.2 & 0 & 0 & 0.2 & 0 \\ 6 & 0 & 2 & -6 & 0 & 4 \end{array}\right] & \begin{array}{c} 4 \\ 5 \\ 6 \\ 1 \\ 2 \\ 3 \end{array} \end{array}$$

$$[K^2] = \begin{array}{c} \\ \end{array} \begin{array}{cccccc} 1 & 2 & 3 & 7 & 8 & 9 \\ \left[\begin{array}{cccccc} 12 & -6 & 0 & -12 & -6 & 0 \\ -6 & 4 & 0 & 6 & 2 & 0 \\ 0 & 0 & 0.2 & 0 & 0 & -0.2 \\ -12 & 6 & 0 & 12 & 6 & 0 \\ -6 & 2 & 0 & 6 & 4 & 0 \\ 0 & 0 & -0.2 & 0 & 0 & 0.2 \end{array}\right] & \begin{array}{c} 1 \\ 2 \\ 3 \\ 7 \\ 8 \\ 9 \end{array} \end{array}$$

Step 4: Develop structure stiffness matrix $[K]$

$$[K] = EI \begin{array}{c} \\ \end{array} \begin{array}{ccccccccc} 1 & 2 & 3 & 4 & 5 & 6 & 7 & 8 & 9 \\ \left[\begin{array}{ccc|cccccc} 24 & -6 & -6 & -12 & 0 & -6 & -12 & -6 & 0 \\ -6 & 4.2 & 0 & 0 & -0.2 & 0 & 6 & 2 & 0 \\ -6 & 0 & 4.2 & 6 & 0 & 2 & 0 & 0 & -0.2 \\ \hline -12 & 0 & 6 & 12 & 0 & 6 & 0 & 0 & 0 \\ 0 & -0.2 & 0 & 0 & 0.2 & 0 & 0 & 0 & 0 \\ -6 & 0 & 2 & 6 & 0 & 4 & 0 & 0 & 0 \\ -12 & 6 & 0 & 0 & 0 & 0 & 12 & 6 & 0 \\ -6 & 2 & 0 & 0 & 0 & 0 & 6 & 4 & 0 \\ 0 & 0 & -0.2 & 0 & 0 & 0 & 0 & 0 & 0.2 \end{array}\right] & \begin{array}{c} 1 \\ 2 \\ 3 \\ 4 \\ 5 \\ 6 \\ 7 \\ 8 \\ 9 \end{array} \end{array}$$

Step 5: Find $\{D_A\}$

$$\{D_R\} = \{0\}$$

$$\{D_A\} = [K_{AA}]^{-1}\{P_A\} = \left\{\begin{array}{c} D_1 \\ D_2 \\ D_3 \end{array}\right\} = \frac{1}{EI}\left\{\begin{array}{c} -1.4583333 \\ -2.0833333 \\ -2.0833333 \end{array}\right\}$$

Step 6: Find $\{P_R\}$

$$\{P_R\} = [K_{RA}]\{D_A\} = \left\{\begin{array}{c} 5 \\ 0.416667 \\ 4.583333 \\ 5 \\ 4.58333 \\ 0.416667 \end{array}\right\}$$

Step 7: Determine element forces

$$\{D^1\} = \begin{Bmatrix} D_4 \\ D_5 \\ D_6 \\ D_1 \\ D_2 \\ D_3 \end{Bmatrix} = \frac{1}{EI} \begin{Bmatrix} 0 \\ 0 \\ 0 \\ -1.4583333 \\ -2.0833333 \\ -2.0833333 \end{Bmatrix}; \{D^2\} = \begin{Bmatrix} D_1 \\ D_2 \\ D_3 \\ D_7 \\ D_8 \\ D_9 \end{Bmatrix} = \frac{1}{EI} \begin{Bmatrix} -1.4583333 \\ -2.0833333 \\ -2.0833333 \\ 0 \\ 0 \\ 0 \end{Bmatrix};$$

$$\{p^1\} = [k^1][T^1]\{D^1\} = \begin{Bmatrix} 5 \\ 0.41667 \\ 4.58333 \\ -5 \\ -0.41667 \\ 0.41667 \end{Bmatrix}$$

$$\{p^2\} = [k^2][T^2]\{D^2\} = \begin{Bmatrix} -5 \\ -0.41667 \\ -0.41667 \\ 5 \\ 0.41667 \\ -4.58333 \end{Bmatrix}$$

The FBD of the elements are shown in Figure 5.13e.

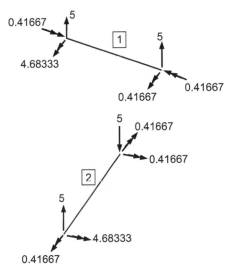

FIGURE 5.13e
FBD of elements (force in kN, m).

Example 5.6

Analyze the space frame shown in Figure 5.14a using the direct stiffness method. All members have a rectangular cross-section of width 10 mm and depth 20 mm. The properties of the material are $E = 2 \times 10^5 \, \text{N/mm}^2$ and $G = 0.8 \times 10^5 \, \text{N/mm}^2$.

Solution

Step 1: Develop the discretized structure and identify the global coordinates.

The discretized structure and global coordinates are shown in Figures 5.14b and c. The global coordinates 1 to 6 are active coordinates and remaining coordinates in Figure 5.13c are restrained coordinates.

Step 2: Identify the force and displacement vectors of the structure.

$$\{P_A\} = \begin{Bmatrix} P_1 \\ P_2 \\ P_3 \\ P_4 \\ P_5 \\ P_6 \end{Bmatrix} = \begin{Bmatrix} 0 \\ 0 \\ 0 \\ 0 \\ 10^6 \, \text{Nmm} \\ 0 \end{Bmatrix}; \{D_A\} = \begin{Bmatrix} D_1 \\ D_2 \\ D_3 \\ D_4 \\ D_5 \\ D_6 \end{Bmatrix}$$

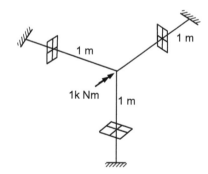

FIGURE 5.14a
Space frame.

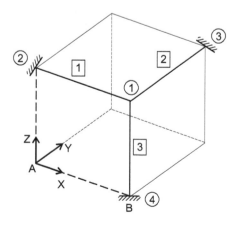

FIGURE 5.14b
Discretized structure.

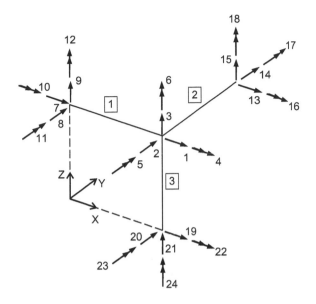

FIGURE 5.14c
Global coordinates.

$$\{P_R\}=\left\{\begin{array}{c} P_7 \\ P_8 \\ \vdots \\ P_{24} \end{array}\right\};\{D_R\}=\left\{\begin{array}{c} D_7 \\ D_8 \\ \vdots \\ D_{24} \end{array}\right\}=\left\{\begin{array}{c} 0 \\ 0 \\ \vdots \\ 0 \end{array}\right\}$$

Step 3: Develop element stiffness matrix.

The nodal coordinates and element connectivity data are given in Tables 5.10 and 5.11. The K node of the elements and its coordinates are given in Table 5.12. The local coordinates of the elements are shown in Figure 5.14d.

TABLE 5.10

Coordinates of Node

Node	X (mm)	Y (mm)	Z (mm)
1	1000	0	1000
2	0	0	1000
3	1000	1000	1000
4	1000	0	0

TABLE 5.11

Element Connectivity Data

Element	Near Node N	Far Node F
1	1	2
2	1	3
3	1	4

TABLE 5.12

Coordinates of K Node

Element	K^a node	Coordinates		
		X (mm)	Y (mm)	Z (mm)
1	A	0	0	0
2	B	1000	0	0
3	A	0	0	0

[a] Points A and B are indicated in Figure 5.14b.

- Element stiffness matrix in local coordinates $[k^i]$
  Section properties:

$$A = 10 \times 20 = 200 \text{ mm}^2$$

$$I_z = \frac{10 \times 20^3}{12} = 6666.67 \text{ mm}^4$$

$$I_y = \frac{20 \times 10^3}{12} = 1666.67 \text{ mm}^4$$

$$I_y = \frac{20 \times 10^3}{12} = 1666.67 \text{ mm}^4; J = \frac{ab^3}{3}\left[1 - 0.63\left(\frac{b}{a}\right) + \frac{0.63}{12}\left(\frac{b}{a}\right)^5\right]; \quad b < a$$

Here, $b = 10$ mm, $a = 20$ mm $\Rightarrow J = 4577.6$ mm^4
Using Eq. 5.17f,

$$[k^1] = [k^2] = [k^3] = \begin{bmatrix} 4000 & 0 & 0 & 0 & 0 & 0 \\ 0 & 16 & 0 & 0 & 0 & 8000 \\ 0 & 0 & 4 & 0 & -2000 & 0 \\ 0 & 0 & 0 & 366208 & 0 & 0 \\ 0 & 0 & -2000 & 0 & 1333336 & 0 \\ 0 & 8000 & 0 & 0 & 0 & 5333336 \\ -4000 & 0 & 0 & 0 & 0 & 0 \\ 0 & -16 & 0 & 0 & 0 & -8000 \\ 0 & 0 & -4 & 0 & 2000 & 0 \\ 0 & 0 & 0 & -366208 & 0 & 0 \\ 0 & 0 & -2000 & 0 & 666668 & 0 \\ 0 & 8000 & 0 & 0 & 0 & 2666668 \end{bmatrix}$$

$$\begin{bmatrix} -40000 & 0 & 0 & 0 & 0 & 0 \\ 0 & -16 & 0 & 0 & 0 & 8000 \\ 0 & 0 & -4 & 0 & -2000 & 0 \\ 0 & 0 & 0 & -366208 & 0 & 0 \\ 0 & 0 & 2000 & 0 & 666668 & 0 \\ 0 & -8000 & 0 & 0 & 0 & 2666668 \\ 40000 & 0 & 0 & 0 & 0 & 0 \\ 0 & 0 & 0 & 0 & 0 & -8000 \\ 0 & 0 & 0 & 0 & 2000 & 0 \\ 0 & 0 & 0 & 366208 & 0 & 0 \\ 0 & 0 & 2000 & 0 & 1333336 & 0 \\ 0 & -8000 & 0 & 0 & 0 & 5333336 \end{bmatrix}$$

- Element transformation matrix $[T^i]$

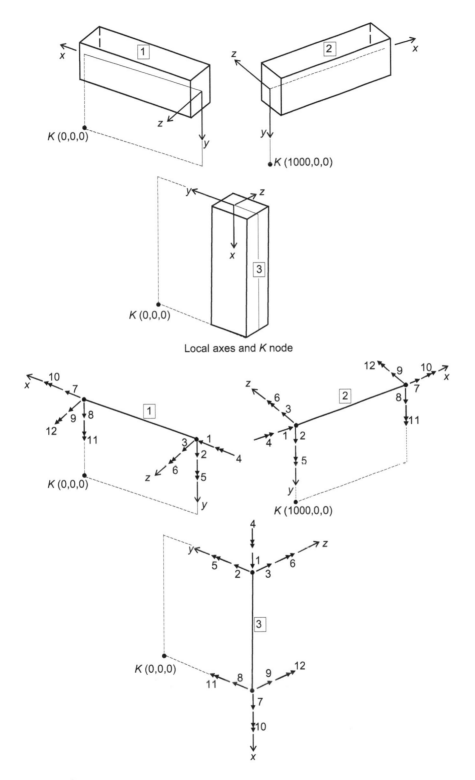

FIGURE 5.14d
Local axes and coordinates (superscripts on coordinates are avoided for clarity in the drawing).

Element 1

$$X_x = X_F - X_N = 0 - 1000 = -1000; X_y = Y_F - Y_N = 0 - 0 = 0; X_z = Z_F - Z_N = 1000 - 1000 = 0$$

$$K_x = X_K - X_N = 0 - 1000 = -1000; K_y = Y_K - Y_N = 0 - 0 = 0;$$

$$K_z = Z_K - Z_N = 0 - 1000 = -1000$$

$$Z_x = X_y K_z - X_z K_y = 0; Z_y = X_z K_x - X_x K_z = -(1000)^2; Z_z = X_x K_y - X_y K_x = 0$$

$$Y_x = Z_y X_z - Z_z X_y = 0; Y_y = Z_z X_x - Z_x X_z = 0; Y_z = Z_x X_y - Z_y X_x = -(1000)^3$$

$$X = \sqrt{X_x^2 + X_y^2 + X_z^2} = 1000; Y = \sqrt{Y_x^2 + Y_y^2 + Y_z^2} = (1000)^3;$$

$$Z = \sqrt{Z_x^2 + Z_y^2 + Z_z^2} = 1000^2$$

Using Eq. 5.17,

$$l_1 = X_x/X = -1; m_1 = X_y/X = 0; n_1 = X_z/X = 0$$

$$l_2 = Y_x/Y = 0; m_2 = Y_y/Y = 0; n_2 = Y_z/Y = -1$$

$$l_3 = Z_x/Z = 0; m_3 = Z_y/Z = -1; n_3 = Z_z/Z = 0$$

By Eq. 5.17e,

$$[T^1] = \begin{bmatrix}
-1 & 0 & 0 & 0 & 0 & 0 & 0 & 0 & 0 & 0 & 0 & 0 \\
0 & 0 & -1 & 0 & 0 & 0 & 0 & 0 & 0 & 0 & 0 & 0 \\
0 & -1 & 0 & 0 & 0 & 0 & 0 & 0 & 0 & 0 & 0 & 0 \\
0 & 0 & 0 & -1 & 0 & 0 & 0 & 0 & 0 & 0 & 0 & 0 \\
0 & 0 & 0 & 0 & 0 & -1 & 0 & 0 & 0 & 0 & 0 & 0 \\
0 & 0 & 0 & 0 & -1 & 0 & 0 & 0 & 0 & 0 & 0 & 0 \\
0 & 0 & 0 & 0 & 0 & 0 & -1 & 0 & 0 & 0 & 0 & 0 \\
0 & 0 & 0 & 0 & 0 & 0 & 0 & 0 & -1 & 0 & 0 & 0 \\
0 & 0 & 0 & 0 & 0 & 0 & 0 & -1 & 0 & 0 & 0 & 0 \\
0 & 0 & 0 & 0 & 0 & 0 & 0 & 0 & 0 & -1 & 0 & 0 \\
0 & 0 & 0 & 0 & 0 & 0 & 0 & 0 & 0 & 0 & 0 & -1 \\
0 & 0 & 0 & 0 & 0 & 0 & 0 & 0 & 0 & 0 & -1 & 0
\end{bmatrix}$$

Element 2

$l_1 = 0, m_1 = 1, n_1 = 0$
$l_2 = 0; m_2 = 0; n_2 = -1$
$l_3 = -1; m_3 = 0; n_3 = 0$

$$[T^2] = \begin{bmatrix}
0 & 1 & 0 & 0 & 0 & 0 & 0 & 0 & 0 & 0 & 0 & 0 \\
0 & 0 & -1 & 0 & 0 & 0 & 0 & 0 & 0 & 0 & 0 & 0 \\
-1 & 0 & 0 & 0 & 0 & 0 & 0 & 0 & 0 & 0 & 0 & 0 \\
0 & 0 & 0 & 0 & 1 & 0 & 0 & 0 & 0 & 0 & 0 & 0 \\
0 & 0 & 0 & 0 & 0 & -1 & 0 & 0 & 0 & 0 & 0 & 0 \\
0 & 0 & 0 & -1 & 0 & 0 & 0 & 0 & 0 & 0 & 0 & 0 \\
0 & 0 & 0 & 0 & 0 & 0 & 0 & 1 & 0 & 0 & 0 & 0 \\
0 & 0 & 0 & 0 & 0 & 0 & 0 & 0 & -1 & 0 & 0 & 0 \\
0 & 0 & 0 & 0 & 0 & 0 & -1 & 0 & 0 & 0 & 0 & 0 \\
0 & 0 & 0 & 0 & 0 & 0 & 0 & 0 & 0 & 0 & 1 & 0 \\
0 & 0 & 0 & 0 & 0 & 0 & 0 & 0 & 0 & 0 & 0 & -1 \\
0 & 0 & 0 & 0 & 0 & 0 & 0 & 0 & 0 & -1 & 0 & 0
\end{bmatrix}$$

Element 3

$l_1 = 0, m_1 = 0, n_1 = -1$
$l_2 = -1; m_2 = 0; n_2 = 0$
$l_3 = 0; m_3 = 1; n_3 = 0$

$$[T^3] = \begin{bmatrix}
0 & 0 & -1 & 0 & 0 & 0 & 0 & 0 & 0 & 0 & 0 & 0 \\
-1 & 0 & 0 & 0 & 0 & 0 & 0 & 0 & 0 & 0 & 0 & 0 \\
0 & 1 & 0 & 0 & 0 & 0 & 0 & 0 & 0 & 0 & 0 & 0 \\
0 & 0 & 0 & 0 & 0 & -1 & 0 & 0 & 0 & 0 & 0 & 0 \\
0 & 0 & 0 & -1 & 0 & 0 & 0 & 0 & 0 & 0 & 0 & 0 \\
0 & 0 & 0 & 0 & 1 & 0 & 0 & 0 & 0 & 0 & 0 & 0 \\
0 & 0 & 0 & 0 & 0 & 0 & 0 & 0 & -1 & 0 & 0 & 0 \\
0 & 0 & 0 & 0 & 0 & 0 & -1 & 0 & 0 & 0 & 0 & 0 \\
0 & 0 & 0 & 0 & 0 & 0 & 0 & 1 & 0 & 0 & 0 & 0 \\
0 & 0 & 0 & 0 & 0 & 0 & 0 & 0 & 0 & 0 & 0 & -1 \\
0 & 0 & 0 & 0 & 0 & 0 & 0 & 0 & 0 & -1 & 0 & 0 \\
0 & 0 & 0 & 0 & 0 & 0 & 0 & 0 & 0 & 0 & 1 & 0
\end{bmatrix}$$

- Element stiffness matrix in global coordinates

	1	2	3	4	5	6
	40000	0	0	0	0	0
	0	0	0	0	0	−2000
	0	0	0	0	8000	0
	0	0	0	366208	0	0
	0	0	8000	0	5333336	0
$[K^1] =$	0	−2000	0	0	0	1333336
	−40000	0	0	0	0	0
	0	−4	0	0	0	2000
	0	0	−16	0	−8000	0
	0	0	0	−366208	0	0
	0	0	8000	0	2666668	0
	0	−2000	0	0	0	666668

7	8	9	10	11	12	
−40000	0	0	0	0	0	1
0	−4	0	0	0	−2000	2
0	0	−16	0	8000	0	3
0	0	0	−366208	0	0	4
0	0	−8000	0	2666668	0	5
0	2000	0	0	0	666668	6
40000	0	0	0	0	0	7
0	4	0	0	0	2000	8
0	0	16	0	−8000	0	9
0	0	0	366208	0	0	10
0	0	−8000	0	5333336	0	11
0	2000	0	0	0	1333336	12

$$[K^2]=$$

	1	2	3	4	5	6
	4	0	0	0	0	-2000
	0	40000	0	0	0	0
	0	0	16	8000	0	0
	0	0	8000	5333336	0	0
	0	0	0	0	366208	0
	-2000	0	0	0	0	1333336
	-4	0	0	0	0	2000
	0	-40000	0	0	0	0
	0	0	-16	-8000	0	0
	0	0	8000	2666668	0	0
	0	0	0	0	-366208	0
	-2000	0	0	0	0	666668

13	14	15	16	17	18	
-4	0	0	0	0	-2000	1
0	-40000	0	0	0	0	2
0	0	-16	8000	0	0	3
0	0	-8000	2666668	0	0	4
0	0	0	0	-366208	0	5
2000	0	0	0	0	666668	6
4	0	0	0	0	2000	13
0	40000	0	0	0	0	14
0	0	16	-8000	0	0	15
0	0	-8000	5333336	0	0	16
0	0	0	0	366208	0	17
2000	0	0	0	0	1333336	18

$$[K^3]=$$

	1	2	3	4	5	6
	16	0	0	0	-8000	0
	0	4	0	2000	0	0
	0	0	4000	0	0	0
	0	2000	0	1333336	0	0
	-8000	0	0	0	5333336	0
	0	0	0	0	0	366208
	-16	0	0	0	8000	0
	0	-4	0	-2000	0	0
	0	0	-40000	0	0	0
	0	2000	0	666668	0	0
	-8000	0	0	0	2666668	0
	0	0	0	0	0	-366208

19	20	21	22	23	24	
-16	0	0	0	-8000	0	1
0	-4	0	2000	0	0	2
0	0	-40000	0	0	0	3
0	-2000	0	666668	0	0	4
8000	0	0	0	2666668	0	5
0	0	0	0	0	-366208	6
16	0	0	0	8000	0	19
0	4	0	-2000	0	0	20
0	0	40000	0	0	0	21
0	-2000	0	1333336	0	0	22
8000	0	0	0	5333336	0	23
0	0	0	0	0	366208	24

Step 4: Develop structure stiffness matrix

$$[K] =$$

	1	2	3	4	5	6	7	8	9	10	11	12	13	14	15	16	17	18	19	20	21	22	23	24
1	40020	0	0	-8000	-2000	-40000	0	0	0	0	0	-4	0	0	0	0	-2000	-16	0	0	0	-8000	0	0
2	0	40008	2000	0	-2000	-4	0	4	0	-2000	0	-40000	0	0	0	0	0	0	4	0	2000	0	0	0
3	0	2000	40032	8000	8000	0	0	0	-16	0	8000	0	0	0	-16	8000	0	0	0	-40000	0	0	0	0
4	-8000	0	8000	7032880	1.1E-07	0	0	0	0	0	0	0	0	0	0	2666668	-366208	666668	0	-2000	0	666668	2666668	-366208
5	-2000	-2000	8000	1.1E-07	3032880	0	0	2000	-8000	0	2666668	666668	0	666668	0	0	-366208	0	8000	0	0	0	2666668	0
6	-40000	-4	0	0	0	3032880	0	4	0	666668	2000	0	0	666668	0	0	0	2000	0	-2000	0	0	0	-366208
7	0	0	0	0	0	0	40000	4	-8000	366208	-8000	0	0	0	0	0	0	0	0	0	0	0	0	0
8	0	4	0	0	2000	4	4	2000	0	0	0	0	0	0	0	0	0	0	0	0	0	0	0	0
9	0	0	-16	0	-8000	0	-8000	0	16	0	5333336	0	0	0	0	0	0	0	0	0	0	0	0	0
10	0	-2000	0	0	0	666668	366208	0	0	0	0	1333336	0	0	0	0	0	0	0	0	0	0	0	0
11	0	0	8000	0	2666668	2000	-8000	0	5333336	0	0	0	0	0	0	0	0	0	0	0	0	0	0	0
12	-4	-40000	0	0	666668	0	0	0	0	1333336	0	40000	0	0	0	0	0	0	0	0	0	0	0	0
13	0	0	0	0	0	0	0	0	0	0	0	0	40000	0	-8000	0	0	0	0	0	0	0	0	0
14	0	0	0	0	666668	666668	0	0	0	0	0	0	0	40000	-8000	5333336	0	0	0	0	0	0	0	0
15	0	0	-16	0	0	0	0	0	0	0	0	0	-8000	-8000	16	0	0	0	0	0	0	0	0	0
16	0	0	8000	2666668	0	0	0	0	0	0	0	0	0	5333336	0	5333336	0	0	0	0	0	0	0	0
17	-2000	0	0	-366208	-366208	0	0	0	0	0	0	0	0	0	0	0	366208	0	0	0	0	0	0	0
18	-16	0	0	666668	0	2000	0	0	0	0	0	0	0	0	0	0	0	1333336	0	0	0	0	0	0
19	0	4	0	0	8000	0	0	0	0	0	0	0	0	0	0	0	0	0	16	0	0	0	8000	0
20	0	0	-40000	-2000	0	-2000	0	0	0	0	0	0	0	0	0	0	0	0	0	4	40000	-2000	0	0
21	0	2000	0	0	0	0	0	0	0	0	0	0	0	0	0	0	0	0	0	40000	40000	0	0	0
22	-8000	0	0	666668	0	0	0	0	0	0	0	0	0	0	0	0	0	0	0	-2000	0	1333336	0	0
23	0	0	0	2666668	2666668	0	0	0	0	0	0	0	0	0	0	0	0	0	8000	0	0	0	5333336	0
24	0	0	0	-366208	0	-366208	0	0	0	0	0	0	0	0	0	0	0	0	0	0	0	0	0	366208

Step 5: Find $\{D_A\}$

$$\{D_R\} = \{0\}$$

$$\{D_A\} = \begin{Bmatrix} D_1 \\ D_2 \\ D_3 \\ D_4 \\ D_5 \\ D_6 \end{Bmatrix} = \begin{Bmatrix} 0.018124425 \\ -4.33067 \times 10^{-7} \\ -0.018122515 \\ 2.06147 \times 10^{-5} \\ 0.090664448 \\ 1.19517 \times 10^{-5} \end{Bmatrix}$$

Step 6: Find $\{P_R\}$ (in N and mm)

$$\{P_R\} = \begin{Bmatrix} -724.977 \\ 0.023905 \\ -725.026 \\ -7.54928 \\ 241627 \\ 7.968663 \\ -0.04859 \\ 0.017323 \\ 0.0125042 \\ -90.0075 \\ -33202 \\ -28.2811 \\ 725.0256 \\ -0.04123 \\ 724.9006 \\ 13.74232 \\ 241627 \\ -4.3768 \end{Bmatrix}$$

Step 7: Determine element forces (in N and mm)

$$\{p^1\} = \begin{bmatrix} -724.98 \\ -725.03 \\ 0.02391 \\ -7.5493 \\ -15.936 \\ -483399 \\ 724.977 \\ 725.026 \\ -0.0239 \\ 7.54928 \\ -7.9687 \\ -241627 \end{bmatrix}$$

$$\{p^2\} = \begin{bmatrix} -0.0173 \\ 0.12504 \\ -0.0486 \\ 33202 \\ 20.3133 \\ 35.0348 \\ 0.01732 \\ -0.125 \\ 0.04859 \\ -33202 \\ 28.2811 \\ 90.0075 \end{bmatrix}$$

$$\{p^3\} = \begin{bmatrix} 724.901 \\ 725.026 \\ 0.04123 \\ -4.3768 \\ -27.486 \\ 483399 \\ -724.9 \\ -725.03 \\ -0.0412 \\ 4.3768 \\ -13.742 \\ 241627 \end{bmatrix}$$

5.8 Introduction to Nonlinear Structural Analysis

All problems are nonlinear in nature. But in many cases, it is sufficient to treat the problems as linear to get reasonably good results. However, in some types of problems, nonlinearity must be considered to get realistic results. For example, to get post yield behavior, post-buckling behavior, etc., nonlinear analysis is required.

There are four sources of nonlinearities in structural analysis:

i. Material nonlinearity—this is due to nonlinear stress–strain relation (constitutive relation) of the material.

ii. Geometric nonlinearity—this type of nonlinearity is due to nonlinear strain–displacement relations.

iii. Force nonlinearity—in these types of nonlinear problems, the direction and magnitude of applied forces changes with deformation.

iv. Kinetic nonlinearity—this is due to change in the displacement boundary conditions with the deformations of the structure.

The basic concepts required for considering material and geometric nonlinearities in the analysis of simple structures are discussed in this section.

5.8.1 Matrix Approach for Nonlinear Analysis

The equilibrium equation for the structure is

$$[K]\{D\} = \{P\}$$

In nonlinear structures[*], the coefficients in $[K]$ will be functions of $\{D\}$. Hence, it is not possible to find the unknown displacement vector explicitly.

The nonlinear problems are solved by means of suitable repetition of linear structural analysis. The two popular numerical approaches for solving nonlinear analysis of structures are as follows: (i) step-by-step application of incremental loads and (ii) regular or modified Newton–Raphson's (N–R) iterative solution under full loads. In both these methods, it is assumed that at each solution step, the analysis is done along a straight line tangent to the force–displacement response of the structure. The equilibrium equation for each step is

$$[K_t]\{\Delta D\} = \{\Delta P\} \tag{5.20}$$

where $[K_t]$ is the tangent stiffness matrix, $\{\Delta D\}$ is the vector of incremental displacements, and $\{\Delta P\}$ is the vector of incremental forces.

5.8.2 Solutions Methods

A nonlinear spring is shown in Figure 5.15. It is required to find the deflection D_A of the spring when it is subjected to a force P_A. The incremental and iterative solution procedures are discussed using this example.

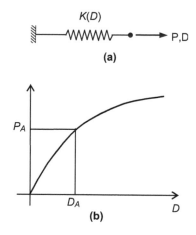

FIGURE 5.15
Nonlinear spring: (a) spring and (b) force–displacement curve.

---

[*] In a general nonlinear problem, both $[K]$ and $\{P\}$ are functions of $\{D\}$. In this section, it is assumed that only $[K]$ is function of $\{D\}$.

5.8.2.1 Incremental Procedure

In this method, the nonlinear problem is approximated as a series of linear problems. The total load is divided into many partial loads or increments. Generally, the load increments need not be of same magnitude. But for simplicity in calculations, usually they are taken as same. The load is applied incrementally. At a particular load increment $\{\Delta P\}$, using the assumed tangent stiffness matrix $[K_t]$, the incremental displacement vector $\{\Delta D\}$ is calculated as

$$\{\Delta D\} = [K_t]^{-1}\{\Delta P\}$$

This procedure is repeated till the total load is reached. The cumulative sum of $\{\Delta D\}$ for all the load steps will be the total displacement.

For the analysis of nonlinear spring shown in Figure 5.15, the total load P_A is divided into n equal load steps. Hence, the incremental load applied at any stage j is $\Delta P_j = P_A/n$. After the application of the ith load increment, the load is

$$P_i = \sum_{j=1}^{i} \Delta P_j$$

and the displacement is

$$D_i = \sum_{j=1}^{n} \Delta D_j$$

The tangent stiffness calculated at the end of previous load step K_{ti-1} is used for calculating ΔD_i, i.e.,

$$\Delta D_i = K_{ti-1}^{-1}\Delta P_i \quad \text{for } i=1 \text{ to } n$$

In the case of the first load increment ($i = 1$), the initial tangent stiffness K_{t0} is used in the above equation.

The incremental procedure is schematically shown in Figure 5.16. The accuracy of this procedure can be increased by increasing the number of load steps n. In addition to the final displacement, the response of the structure at any intermediate load is also obtained using this method.

5.8.2.2 Iterative Procedure

In the iterative procedure, the full load $\{P\}$ is applied on the structure in the beginning. Since the exact value of stiffness of the structure is not known initially, an approximate value of stiffness is sued for analysis. Hence, the calculated displacement will not be equal to its actual value. The force developed in the structure for the calculated displacement is determined. This force is in fact the force developed in the structure to maintain equilibrium for the calculated displacement and is denoted as $\{P_e\}$. Now, the unbalanced force vector $\{\Delta P\} = \{P\} - \{P_e\}$ is applied on the structure and the incremental displacement vector is determined. The cumulative sum of the displacements is found and the new value of $\{P_e\}$ is calculated for this displacement. The procedure is repeated till $\{P_e\}$ is equal to $\{P\}$ or till the difference between $\{P\}$ and $\{P_e\}$ reaches an acceptable error criterion.

For the ith iteration, the load required for analysis is

$$\{\Delta P_i\} = \{P\} - \{P_{ei-1}\} \tag{5.21}$$

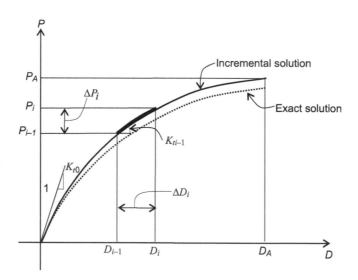

FIGURE 5.16
Incremental procedure.

where $\{P_{ei-1}\}$ is the load equilibrated after the previous iteration (i.e., at $(i-1)$ iteration). The increment in displacement vector $\{\Delta D_i\}$ is calculated as

$$\{\Delta D_i\} = [K_t^i]^{-1}\{\Delta P_i\} \tag{5.22}$$

where $[K_t^i]$ is the tangent stiffness matrix used in the ith iteration. The total displacement after i^{th} iteration is

$$\{D_i\} = \sum_{j=1}^{n}\{\Delta D_j\} \tag{5.23}$$

In Eq. 5.22, if the tangent stiffness matrix at the end of the previous iteration step $[K_{ti-1}]$ is used as $[K_t^i]$, then the procedure is called Newton–Raphson's (N–R) method. In this case, Eq. 5.22 can be written as

$$\{\Delta D_j\} = [K_{ti-1}]\{\Delta P_i\}$$

Alternatively, the initial tangent stiffness matrix $[K_{t0}]$ can be used as $[K_t^i]$ for all iterations. This procedure is known as modified N–R method. Equation 5.22 for this approach is

$$\{\Delta D_i\} = [K_{t0}]^{-1}\{\Delta P_i\}$$

Since there is no need to update the tangent stiffness matrix, there will be reduction in computational steps at a particular iteration if modified N–R method is used. But the number of iterations required will be more than that needed for N–R method.

Unlike in the incremental method, where the response of the structure at intermediate load steps is obtained, in the iterative method, the response of the structure only at the final load can be determined. The principal disadvantage of the iterative method is that there is no guarantee that the results from analysis will converge to the exact solution. However, the procedure is simple and will take less computational time when compared with the incremental method.

To find the displacement in the nonlinear spring shown in Figure 5.14 using N–R method, the load P_A is applied on the spring. Hence, $\Delta P_1 = P_A$ and $K_t^1 = K_{t0}$, the initial tangent stiffness of the spring. Equation 5.22 is used to find the incremental displacement ΔD_1. The total displacement at the end of the first iteration is $D_1 = \Delta D_1$.

The force developed in the spring for the displacement D_1 is P_{e1}. The unbalanced force $P_A - P_{e1}$ is calculated and is applied on the spring and the incremental displacement ΔD_2 is calculated. Thus, $\Delta P_2 = P_A - P_{e1}$ and $K_t^2 = K_{t1}$. The displacement at the end of the second iteration is $D_2 = D_1 + \Delta D_2 = \Delta D_1 + \Delta D_2$.

The above steps are repeated till the unbalanced force becomes negligibly small. To get the solution using modified N–R method, the above steps are repeated by using $K_t^1 = K_{t0}$. The two procedures are graphically shown in Figure 5.17.

5.8.3 Material Nonlinearity

Consider a rod subjected to axial loading shown in Figure 5.18a. It is required to find the displacement of the rod up to a load of 1800 kN. The stress–strain relation of the material in the bar is shown in Figure 5.18b. The steps for finding the displacement are given below.

- Stress due to P

$$\text{Cross sectional area } A = 70 \times 70 = 4900 \text{ mm}^2$$

$$\text{Stress } \sigma = P/A = \frac{1800 \times 10^3}{4900} = 367.35 \text{ N/mm}^2 \left(> \text{yield stress } \sigma_y \right)$$

- Displacement D due to P

$$\sigma = 367.35 \text{ N/mm}^2$$

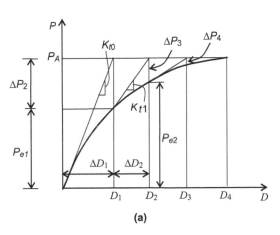

(a)

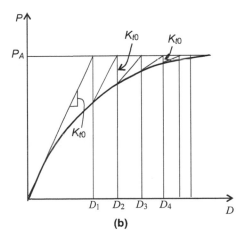

(b)

FIGURE 5.17
Iterative procedure: (a) N–R procedure and (b) modified N–R procedure.

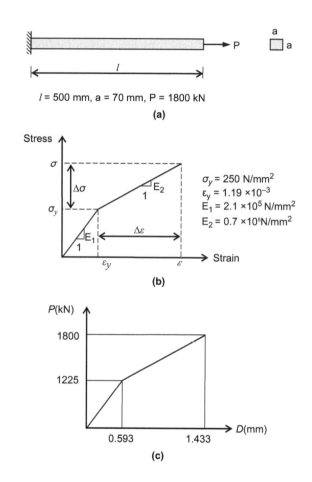

FIGURE 5.18
Effects of material nonlinearity in structural analysis: (a) rod subjected to axial load, (b) stress–strain curve, and (c) force–elongation curve.

From Figure 5.18b,

$$\Delta\sigma = \sigma - \sigma_y = 367.35 - 250 = 117.35 \text{ N/mm}^2$$

$$\Delta\varepsilon = \varepsilon - \varepsilon_y = \frac{\Delta\sigma}{E_2} = \frac{117.35}{0.7 \times 10^5} = 1.6764 \times 10^{-3}$$

$$\Rightarrow \varepsilon = \Delta\varepsilon + \varepsilon_y = 2.867 \times 10^{-3}$$

$$\varepsilon = \frac{D}{l} \Rightarrow D = \varepsilon l = 2.867 \times 10^{-3} \times 500 = 1.433 \text{ mm}$$

- Displacement D_y at yield load P_y

$$P_y = \sigma_y A = 1226 \text{ kN}$$

$$D_y = \varepsilon_y l = 1.19 \times 10^{-3} \times 500 = 0.593 \text{ mm}$$

The force elongation curve of the rod up to a force of 1800 kN is shown in Figure 5.18c.

The above problem is a simple example to illustrate the effects of material nonlinearity in structures. From the example, it is observed that a nonlinear constitutive relation of the material leads to a nonlinear response of the structure.

Example 5.7

Find the elongation of the rod shown in Figure 5.18a using the matrix stiffness method.

Solution

The rod is modeled using a single truss element shown in Figure 5.19. The element, nodes, and the global coordinate are indicated in the figure. In order to reduce the hand computation time, only active coordinate is considered.

The tangent stiffness matrix of the structure is

$$K_t = \frac{AE_t}{l}$$

$$\text{where } E_t = \begin{cases} E_1, \varepsilon \le \varepsilon_y \\ E_2, \varepsilon > \varepsilon_y \end{cases}$$

Method 1: Incremental Procedure

Let the number of increments $n = 5$

$$\text{Load increment } \Delta P = \frac{P}{n} = \frac{1800}{5} = 360 \text{ kN}$$

For the ith load increment,

$$\Delta D_i = \frac{\Delta P_i}{K_{ti-1}}$$

- Load increment 1

$$\Delta P_1 = 360 \text{ kN}$$

$$K_{t0} = \frac{AE_1}{l} = \frac{70 \times 70 \times 2.1 \times 10^5}{500} = 2058000 \text{ N/mm}$$

$$\Delta D_1 = \frac{\Delta P_1}{K_{t0}} = \frac{360 \times 10^3}{2058000} = 0.1749 \text{ mm}$$

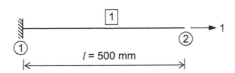

FIGURE 5.19
Discretized structure.

$$P_1 = \Delta P_1 = 360 \text{ kN}$$

$$D_1 = \Delta D_1 = 0.1748 \text{ mm}$$

$$\text{Strain } \varepsilon = \frac{D}{l} = \frac{0.1749}{500} = 3.498 \times 10^{-4} < \varepsilon_y$$

$$\text{Hence, } K_{t1} = \frac{AE_1}{l} = 2058000 \text{ N/mm}$$

- Load increment 2

$$\Delta P_2 = 360 \text{ kN}$$

$$K_{t1} = 2058000 \text{ N/mm}$$

$$\Delta D_2 = \frac{\Delta P_2}{K_{t1}} = 0.1749 \text{ mm}$$

$$P_2 = P_1 + \Delta P_2 = \Delta P_1 + \Delta P_2 = 720 \text{ kN}$$

$$D_2 = D_1 + \Delta D_2 = \Delta D_1 + \Delta D_2 = 0.3499 \text{ mm}$$

$$\varepsilon = \frac{0.3499}{500} = 6.996 \times 10^{-4} < \varepsilon_y$$

$$\Rightarrow K_{t2} = \frac{AE_1}{l} = 2058000 \text{ N/mm}$$

- Load increment 3

$$\Delta P_3 = 360 \text{ kN}$$

$$K_{t2} = 2058000 \text{ N/mm}$$

$$\Delta D_3 = \frac{\Delta P_3}{K_{t2}} = 0.1749 \text{ mm}$$

$$P_3 = P_2 + \Delta P_3 = 1080 \text{ kN}$$

$$D_3 = D_2 + \Delta D_3 = 0.5248 \text{ mm}$$

$$\varepsilon = \frac{0.5248}{500} = 1.0496 \times 10^{-3} < \varepsilon_y$$

$$\Rightarrow K_{t3} = \frac{AE_1}{l} = 2058000 \text{ N/mm}$$

- Load increment 4

$$\Delta P_4 = 360 \text{ kN}$$

$$K_{t3} = 2058000 \text{ N/mm}$$

$$\Delta D_4 = \frac{\Delta P_4}{K_{t3}} = 0.1749 \text{ mm}$$

$$P_4 = P_3 + DP_4 = 1440 \text{ kN}$$

$$D_4 = D_3 + \Delta D_4 = 0.6997 \text{ mm}$$

$$\varepsilon = \frac{0.6997}{500} = 1.3994 \times 10^{-3} > \varepsilon_y$$

$$\Rightarrow K_{t4} = \frac{AE_2}{l} = \frac{70 \times 70 \times 0.7 \times 10^5}{500} = 686000 \text{ N/mm}$$

- Load increment 5

$$\Delta P_5 = 360 \text{ kN}$$

$$K_{t4} = 686000 \text{ N/mm}$$

$$\Delta D_5 = \frac{\Delta P_5}{K_{t4}} = 0.5248 \text{ mm}$$

At the end of five load increments,

$$P_5 = P_4 + \Delta P_5 = 1800 \text{ kN} = P$$

$$D_5 = D_4 + \Delta D_5 = 1.2245 \text{ mm}$$

Hence, the displacement is 1.2245 mm. The exact value of displacement is 1.433 mm. The accuracy can be increased by increasing the number of load increments n. If n is taken as 50, then the displacement obtained is 1.4 mm.

Method 2: Iterative Procedure Using N–R method

Iteration 1

$$P_{e2} = \sigma \times A = 1799.99 \text{ kN}$$

$$\Delta P_1 = 1800 \text{ kN}$$

$$K_{t0} = 2058000 \text{ N/mm}$$

$$\Delta D_1 = \frac{\Delta P_1}{K_{t0}} = 0.87464 \text{ mm}$$

$$D_1 = \Delta D_1 = 0.87464 \text{ mm}$$

$$\varepsilon = \frac{0.87464}{500} = 1.7493 \times 10^{-3} > \varepsilon_y$$

$$\Rightarrow K_{t1} = \frac{AE_2}{l} = 686000 \text{ N/mm}$$

From Figure 5.18b,

For $\varepsilon = 1.7493 \times 10^{-3}$, $\sigma = \sigma_y + (\varepsilon - \varepsilon_y) \, E_2 = 281.12 \text{ N/mm}^2$

Force developed in the rod due to

$$D_1 = P_{e1} = \sigma \times A = \frac{28.12 \times 70 \times 70}{1000} = 1416.67 \text{ kN}$$

Iteration 2

$$\Delta P_2 = P - P_{e1} = 1800 - 1416.67 = 383.33 \text{ kN}$$

$$\Delta D_2 = \frac{\Delta P_2}{K_{t1}} = 0.55879 \text{ mm}$$

$$D_2 = D_1 + \Delta D_2 = 1.4334 \text{ mm}$$

$$\varepsilon = \frac{1.4334}{500} = 2.86686 \times 10^{-3}$$

The corresponding stress $\sigma = 367.35 \text{ N/mm}^2$

$$P_{e2} = \sigma \times A = 1799.99 \text{ kN}$$

The difference between P and P_{e2} is very small and hence the iteration is stopped at this stage. The displacement obtained using NR method is 1.4334 mm which is very close to the exact value (1.4333 mm).

5.8.4 Geometric Nonlinearity

Geometric nonlinearity is caused by the large deflection occurring in the structure. Due to this, the equilibrium equations should be developed based on the deformed configuration of the structure. The need for considering geometric nonlinearity in structural analysis is explained using the structure shown in Figure 5.20a.

In the structure, the two elements AB and BC are hinged at the ends and the forces acts at the hinges. Hence, only axial force develops in the elements. Due to this, if the equilibrium equation is written based on the undeformed configuration of the structure (Figure 5.20a), it will lead to the conclusion that vertical reaction cannot be developed at the supports. In other words, if linear analysis is used (i.e., analysis based on small deflection assumption), then the analysis indicates that the structure will not be able to take the load P. But in the actual case, the structure is capable of resisting the load P. This example indicates the need for considering the geometric nonlinearity in the analysis of structures. The equilibrium equations must be written using the deformed configuration of the structure (Figure 5.20b).

The unknown deflection of the structure is determined using Castigliano's first theorem[*]. The strain ε in the element AB is

$$\varepsilon = \frac{\Delta l}{l} = \frac{AB' - AB}{AB} = \frac{\sqrt{l^2 + D^2} - l}{l} = \sqrt{1 + \left(\frac{D}{l}\right)^2} - 1$$

$$\Rightarrow \varepsilon = \left[1 + \left(\frac{D}{l}\right)^2\right]^{1/2} - 1 \approx \frac{1}{2}\left(\frac{D}{l}\right)^2$$

---

[*] Castigliano's first theorem is applicable for nonlinear structures.

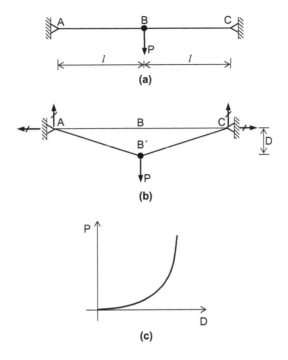

FIGURE 5.20
Illustration of geometric nonlinearity: (a) structure, (b) deformed structure, and (c) force–displacement curve.

The strain energy of the element is

$$U_{AB} = U_{BC} = \frac{1}{2} \int \sigma \varepsilon \, dv$$

If the material is linearly elastic ($\sigma = E\varepsilon$) and if A is the cross-sectional area of the element, then

$$U_{AB} = U_{BC} = \frac{EA}{2} \int_0^l \varepsilon^2 dx$$

The strain energy of the structure is

$$U = U_{AB} + U_{BC} = EA \int_0^l \varepsilon^2 dx = EA \int_0^l \left[\frac{1}{2} \left(\frac{D}{l} \right)^2 \right]^2 dx$$

$$\Rightarrow \quad U = \frac{1}{4} EAl \left(\frac{D}{l} \right)^4$$

By Castigliano's first theorem (Eq. 2.40),

$$P = \frac{\partial U}{\partial D} \Rightarrow P = EA \frac{D^3}{l^3}$$

Thus, the deflection D is

$$D = l\left(\frac{P}{EA}\right)^{1/3}$$

From the above equation, it can be seen that the force–displacement relation (Figure 5.20c) of the structure is nonlinear.

5.8.5 Stiffness Matrix of Truss Elements

The stiffness matrix of truss elements used for geometric nonlinear analysis is derived in this section.

5.8.5.1 Strain–Displacement Relation

A truss element of length dx is shown in Figure 5.21. Under the action of external forces, the truss element gets displaced from AB to A′B′. The displacements at the two nodes along the x- and y-axes are also indicated in the figure.

It is assumed that the displacement from AB to A′B′ occurs in three stages (Figure 5.22). First the truss element gets displaced from AB to A_1B_1, then from A_1B_1 to A_2B_2, and finally from A_2B_2 to A′B′.

The strain in the element ε_x is therefore equal to the sum of the strains developed in the element during these three stages (ε_I, ε_{II}, ε_{III}). Hence,

$$\varepsilon_x = \varepsilon_I + \varepsilon_{II} + \varepsilon_{III}$$

- Strain at Stage I (ε_I)

$$\varepsilon_I = \frac{A_1B_1 - AB}{AB} = \frac{\left[dx + \left(u + \dfrac{du}{dx}dx\right) - u\right] - dx}{dx} = \frac{dx\left(1 + \dfrac{du}{dx}\right) - dx}{dx} = \frac{du}{dx}$$

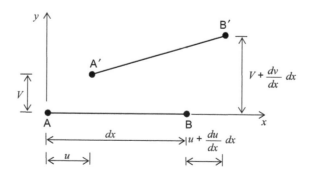

FIGURE 5.21
Truss element.

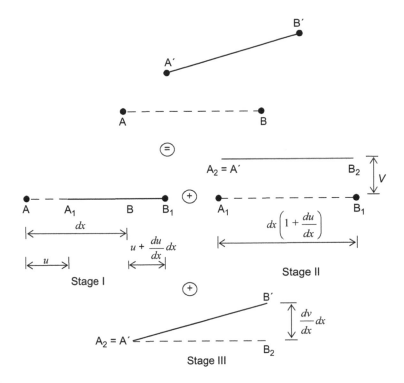

FIGURE 5.22
Displacement of the truss element.

- Strain at Stage II (ε_{II})

$$\varepsilon_{II} = \frac{A_2B_2 - A_1B_1}{A_1B_1} = \frac{dx\left(1+\dfrac{du}{dx}\right) - dx\left(1+\dfrac{du}{dx}\right)}{dx\left(1+\dfrac{du}{dx}\right)} = 0$$

- Strain at Stage III (ε_{III})
  From Figure 5.22,

$$A_2B_2 = \left(1+\frac{du}{dx}\right),\; B_2B' = \frac{dv}{dx}dx$$

$$A'B' = \sqrt{\left(A_2B_2\right)^2 + \left(B_2B'\right)^2} = \sqrt{\left[dx\left(1+\frac{du}{dx}\right)\right]^2 + \left[\frac{dv}{dx}dx\right]^2}$$

$$= dx\left[1+\left(2\frac{du}{dx}+\left(\frac{du}{dx}\right)^2+\left(\frac{dv}{dx}\right)^2\right)\right]^{1/2} \tag{a}$$

The binomial series expansion of the function $(1 + x)^n$ is

$$(1+x)^n = 1 + \frac{nx}{1!} + \frac{n(n-1)x^2}{2!} + \cdots$$ (b)

If $n = \frac{1}{2}$, then

$$(1+x)^{1/2} = 1 + \frac{x}{2} - \frac{x^2}{8} + \cdots$$ (c)

For $n = -1$, then

$$(1+x)^{-1} = 1 - x + x^2 - x^3 + \cdots$$ (d)

Using Eq. (c), Eq. (a) can be written as

$$A'B' = dx\left[1 + \frac{1}{2}\left(2\frac{du}{dx} + \left(\frac{du}{dx}\right)^2 + \left(\frac{dv}{dx}\right)^2\right) - \frac{1}{8}\left(4\left(\frac{du}{dx}\right)^2 + \cdots\right) + \cdots\right]$$

$$\approx dx\left[1 + \frac{du}{dx} + \frac{1}{2}\left(\frac{dv}{dx}\right)^2\right]$$

Therefore, $A'B' - A_2B_2 = dx\left[1 + \frac{du}{dx} + \frac{1}{2}\left(\frac{dv}{dx}\right)^2\right] - dx + \left(1 + \frac{du}{dx}\right) = \frac{1}{2}\left(\frac{dv}{dx}\right)^2 dx$

$$\varepsilon_{III} = \frac{A'B' - A_2B_2}{A_2B_2} = \frac{\frac{1}{2}\left(\frac{dv}{dx}\right)^2 dx}{dx\left(1 + \frac{du}{dx}\right)} = \frac{1}{2}\left[1 + \frac{du}{dx}\right]^{-1}\left(\frac{dv}{dx}\right)^2$$

Using Eq. (d)

$$\left[1 + \frac{du}{dx}\right]^{-1} = 1 - \left(\frac{du}{dx}\right) + \left(\frac{du}{dx}\right)^2 + \cdots$$

$$\Rightarrow \left[1 + \frac{du}{dx}\right]^{-1}\left(\frac{dv}{dx}\right)^2 = \left(\frac{dv}{dx}\right)^2 - \left(\frac{du}{dx}\right)\left(\frac{dv}{dx}\right)^2 + \left(\frac{du}{dx}\right)^2\left(\frac{dv}{dx}\right)^2 + \cdots \approx \left(\frac{dv}{dx}\right)^2$$

$$\therefore \quad \varepsilon_{III} = \frac{1}{2}\left[1 + \frac{du}{dx}\right]^{-1}\left(\frac{dv}{dx}\right)^2 = \frac{1}{2}\left(\frac{dv}{dx}\right)^2$$

Thus,

$$\varepsilon_x = \varepsilon_I + \varepsilon_{II} + \varepsilon_{III} = \frac{du}{dx} + 0 + \frac{1}{2}\left(\frac{dv}{dx}\right)^2$$

$$\Rightarrow \quad \varepsilon_x = \frac{du}{dx} + \frac{1}{2}\left(\frac{dv}{dx}\right)^2$$ (5.24)

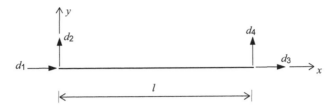

FIGURE 5.23
Truss element and end displacements.

5.8.5.2 Stiffness Matrix

The end displacements of the truss element are shown in Figure 5.23. At any point in the truss element, the displacement functions $u(x)$ and $v(x)$ are assumed to have linear variation. Hence,

$$u(x) = a_1 + a_2 x$$
$$v(x) = b_1 + b_2 x$$

The coefficients are determined using the displacement conditions at the two nodes.
At $x = 0$; $u = d_1$ and $v = d_2$
$x = l$; $u = d_3$ and $v = d_4$
Thus,

$$u = d_1 + \frac{(d_3 - d_1)}{l} x$$

$$v = d_2 + \frac{(d_4 - d_2)}{l} x$$

$$\Rightarrow \quad \frac{du}{dx} = \frac{(d_3 - d_1)}{l} \tag{a}$$

$$\frac{dv}{dx} = \frac{(d_4 - d_2)}{l} \tag{b}$$

The strain energy U stored in the element is

$$U = \frac{1}{2} \int \sigma_x \varepsilon_x dv$$

For linearly elastic material,

$$U = \frac{AE}{2} \int_0^l \varepsilon_x^2 dx$$

Substituting Eq. 5.24 in the above equation gives

$$U = \frac{AE}{2} \int_0^l \left[\frac{du}{dx} + \frac{1}{2}\left(\frac{dv}{dx}\right)^2 \right]^2 dx$$

$$\Rightarrow \quad U = \frac{AE}{2} \int_0^l \left[\left(\frac{du}{dx}\right)^2 + \left(\frac{du}{dx}\right)\left(\frac{dv}{dx}\right)^2 + \frac{1}{4}\left(\frac{dv}{dx}\right)^4 \right] dx$$

Neglecting higher order terms gives

$$U = \frac{AE}{2} \int_0^l \left[\left(\frac{du}{dx}\right)^2 + \left(\frac{du}{dx}\right)\left(\frac{dv}{dx}\right)^2 \right] dx$$

Substituting Eqs. (a) and (b) in the above equation gives

$$U = \frac{AE}{2l^2} \int_0^l \left(d_1^2 - 2d_1d_3 + d_3^2\right) dx + \frac{AE}{2l^3} \int_0^l (d_3 - d_1)\left(d_2^2 - 2d_2d_4 + d_4^2\right) dx$$

$$= \frac{AE}{2l}\left(d_1^2 - 2d_1d_3 + d_3^2\right) + \frac{AE}{2l^2}(d_3 - d_1)\left(d_2^2 - 2d_2d_4 + d_4^2\right)$$

If F is the axial tensile force developed in the truss element, then

$$F = \frac{AE}{l}(d_3 - d_1)$$

Hence,

$$U = \frac{AE}{2l}\left(d_1^2 - 2d_1d_3 + d_3^2\right) + \frac{F}{2l}\left(d_2^2 - 2d_2d_4 + d_4^2\right)$$

Using Castigliano's first theorem, we obtain

$$p_i = \frac{\partial U}{\partial d_i}$$

$$p_1 = \frac{\partial U}{\partial d_1} = \frac{AE}{l}(d_1 - d_3)$$

$$p_2 = \frac{\partial U}{\partial d_2} = \frac{F}{l}(d_2 - d_4)$$

$$p_3 = \frac{\partial U}{\partial d_3} = \frac{AE}{l}(-d_1 + d_3) \qquad\qquad (c)$$

$$p_4 = \frac{\partial U}{\partial d_4} = \frac{F}{l}(-d_2 + d_4)$$

Equation (c) in matrix form gives

$$
\begin{Bmatrix} p_1 \\ p_2 \\ p_3 \\ p_4 \end{Bmatrix} = \frac{AE}{l} \begin{bmatrix} 1 & 0 & -1 & 0 \\ 0 & 0 & 0 & 0 \\ -1 & 0 & 1 & 0 \\ 0 & 0 & 0 & 0 \end{bmatrix} \begin{Bmatrix} d_1 \\ d_2 \\ d_3 \\ d_4 \end{Bmatrix} + \frac{F}{l} \begin{bmatrix} 0 & 0 & 0 & 0 \\ 0 & 1 & 0 & -1 \\ 0 & 0 & 0 & 0 \\ 0 & -1 & 0 & 1 \end{bmatrix} \begin{Bmatrix} d_1 \\ d_2 \\ d_3 \\ d_4 \end{Bmatrix}
$$

or

$$
\{p\} = [k_E]\{d\} + [k_G]\{d\} = [k]\{d\}
$$

The stiffness matrix of the truss element [k] is

$$
[k] = [k_E] + [k_G] \tag{5.25}
$$

[k_E] is the *elastic stiffness matrix* and is equal to

$$
[k_E] = \frac{AE}{l} \begin{bmatrix} 1 & 0 & -1 & 0 \\ 0 & 0 & 0 & 0 \\ -1 & 0 & 1 & 0 \\ 0 & 0 & 0 & 0 \end{bmatrix} \tag{5.26}
$$

The elastic stiffness matrix is the element stiffness matrix of the truss element used in linear analysis.

[k_G] is the *geometrical stiffness matrix* and is given by

$$
[k_G] = \frac{F}{l} \begin{bmatrix} 0 & 0 & 0 & 0 \\ 0 & 1 & 0 & -1 \\ 0 & 0 & 0 & 0 \\ 0 & -1 & 0 & 1 \end{bmatrix} \tag{5.27}
$$

5.8.6 Stiffness Matrix of the Structure

The elastic and geometrical stiffness matrix of the elements is calculated and they are assembled to get the stiffness matrix of the structure [K]. Thus,

$$
[K] = [K_E] + [k_G] \tag{5.28}
$$

Hence, for the analysis of geometrically nonlinear structures, the tangent stiffness matrix [K_t] is

$$
[K_t] = [K_E] + [k_G] \tag{5.29}
$$

Example 5.8

Find the deflection at the loaded point of the cable shown in Figure 5.24a. The span is 25 m and the sag of the cable is 2.5 m.

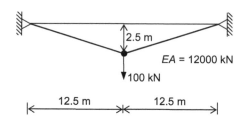

FIGURE 5.24a
Cable structure.

Solution

The cable is modeled using truss elements. The discretized structure, global coordinates, and local coordinates are shown in Figures 5.24b–d. The element connectivity data are given in Table 5.13.

The load vector is

$$\{P_A\} = \left\{ \begin{array}{c} P_1 \\ P_2 \end{array} \right\} = \left\{ \begin{array}{c} 0 \\ -100 \end{array} \right\} \text{kN}$$

Method 1: Incremental Procedure

Let the number of increments $n = 100$

$$\text{Load increment } \{\Delta P_A\} = \frac{1}{100}\{P_A\} = \left\{ \begin{array}{c} 0 \\ -1 \end{array} \right\} \text{kN}$$

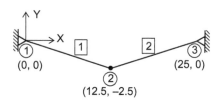

FIGURE 5.24b
Discretized structure.

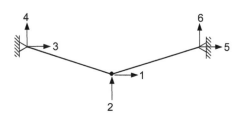

FIGURE 5.24c
Global coordinates.

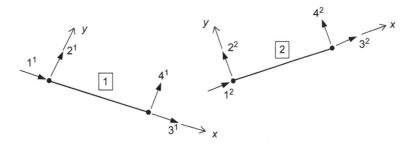

FIGURE 5.24d
Local coordinates.

TABLE 5.13
Element Connectivity Data

Element	Near Node N	Far Node F
1	1	2
2	2	3

For the ith load increment,

$$\{\Delta D_{Ai}\} = [K_{ti-1}]^{-1}\{\Delta P_{Ai}\}$$

Load increment 1

- Increment force vector

$$\{\Delta P_{A1}\} = \left\{\begin{array}{c} 0 \\ -1 \end{array}\right\} \text{kN}$$

- Tangent stiffness matrix of the element in local coordinates

$$AE = 12000 \text{ kN/m}^2$$

$l = \sqrt{12.5^2 + 2.5^2} = 12.7475 = l_0$, the initial length of element.

Initially the truss element is not strained

$$\Rightarrow F = 0$$

$$[k_{E0}] = \frac{AE}{l}\begin{bmatrix} 1 & 0 & -1 & 0 \\ 0 & 0 & 0 & 0 \\ -1 & 0 & 1 & 0 \\ 0 & 0 & 0 & 0 \end{bmatrix} = \begin{bmatrix} 941.3574 & 0 & -941.3574 & 0 \\ 0 & 0 & 0 & 0 \\ -941.3574 & 0 & 941.3574 & 0 \\ 0 & 0 & 0 & 0 \end{bmatrix}$$

$$[k_{G0}] = \frac{F}{l}\begin{bmatrix} 0 & 0 & 0 & 0 \\ 0 & 1 & 0 & -1 \\ 0 & 0 & 0 & 0 \\ 0 & -1 & 0 & 1 \end{bmatrix} = \begin{bmatrix} 0 & 0 & 0 & 0 \\ 0 & 0 & 0 & 0 \\ 0 & 0 & 0 & 0 \\ 0 & 0 & 0 & 0 \end{bmatrix}$$

$$\left[k_{t0}^1\right]=\left[k_{t0}^2\right]=\left[k_{E0}\right]+\left[k_{G0}\right]=\begin{bmatrix} 941.3574 & 0 & -941.3574 & 0 \\ 0 & 0 & 0 & 0 \\ -941.3574 & 0 & 941.3574 & 0 \\ 0 & 0 & 0 & 0 \end{bmatrix}$$

- Element transformation matrix $[T^i]$

Element 1

$$l=\sqrt{(X_F-X_N)^2+(Y_F-Y_N)^2}=\sqrt{(12.5)^2+(-2.5)^2}=12.7475 \text{ m}$$

$$c=\frac{X_F-X_N}{l}=\frac{12.5}{12.7475}=0.980581$$

$$s=\frac{Y_F-Y_N}{l}=\frac{-2.5}{12.7475}=-0.196116$$

$$[T^i]=\begin{bmatrix} c & s & 0 & 0 \\ -s & c & 0 & 0 \\ 0 & 0 & c & s \\ 0 & 0 & -s & c \end{bmatrix}$$

$$\left[T_0^1\right]=\begin{bmatrix} 0.980581 & -0.196116 & 0 & 0 \\ 0.196116 & 0.980581 & 0 & 0 \\ 0 & 0 & 0.980581 & -0.196116 \\ 0 & 0 & 0.196116 & 0.980581 \end{bmatrix}$$

Element 2

$$c=0.980581; s=0.196116$$

$$\left[T_0^2\right]=\begin{bmatrix} 0.980581 & 0.196116 & 0 & 0 \\ -0.196116 & 0.980581 & 0 & 0 \\ 0 & 0 & 0.980581 & 0.196116 \\ 0 & 0 & -0.196116 & 0.980581 \end{bmatrix}$$

- Tangent stiffness matrix of the element in global coordinates

$$\left[K_t^i\right]=\left[T^i\right]^T\left[k_t^i\right]\left[T^i\right]$$

$$\left[K_{t0}^1\right]=\left[T_0^1\right]^T\left[k_{t0}^1\right]\left[T_0^1\right]$$

$$=\begin{array}{cccc} \quad 3 & \quad 4 & \quad 1 & \quad 2 \end{array}$$
$$\begin{bmatrix} 905.1514 & -181.0303 & -905.1514 & 181.0303 \\ -181.0303 & 36.20606 & 181.0303 & -36.20606 \\ -905.1514 & 181.0303 & 905.1514 & -181.0303 \\ 181.0303 & -36.20606 & -181.0303 & 36.20606 \end{bmatrix}\begin{array}{c} 3 \\ 4 \\ 1 \\ 2 \end{array}$$

$$\left[K_{t0}^2\right]=\left[T_0^2\right]^T\left[k_{t0}^2\right]\left[T_0^2\right]^T$$

$$= \begin{bmatrix} 905.1514 & 181.0303 & -905.1514 & -181.0303 \\ 181.0303 & 36.20606 & -181.0303 & -36.20606 \\ -905.1514 & -181.0303 & 905.1514 & 181.0303 \\ -181.0303 & -36.20606 & 181.0303 & 36.20606 \end{bmatrix} \begin{matrix} 1 \\ 2 \\ 5 \\ 6 \end{matrix}$$

(with column headers 1, 2, 5, 6)

- Initial tangent stiffness matrix of the structure

$$[K_{t0}] = \begin{bmatrix} 1810.303 & 0 \\ 0 & 72.41211 \end{bmatrix} \begin{matrix} 1 \\ 2 \end{matrix}$$

(with column headers 1, 2)

Note: The stiffness matrix with respect to active coordinates alone is given above.
- Incremental displacement vector

$$\{\Delta D_{A1}\} = [K_{t0}]^{-1} \{\Delta P_{A1}\} = \left\{ \begin{matrix} 0 \\ -0.01381 \end{matrix} \right\} m$$

After the first load increment,

$$\{P_{A1}\} = \{\Delta P_{A1}\} = \left\{ \begin{matrix} 0 \\ -1 \end{matrix} \right\} kN$$

$$\{D_{A1}\} = \{\Delta D_{A1}\} = \left\{ \begin{matrix} 0 \\ -0.01381 \end{matrix} \right\} m$$

The coordinates of the nodes are shown in Figure 5.24e.

$$\text{Length of element } l = \sqrt{12.5^2 + 2.51381^2} = 12.7503 \text{ m}$$

$$\text{Strain in the element } \varepsilon = \frac{l - l_0}{l_0} = \frac{12.7503 - 12.7475}{12.7475} = 2.1965 \times 10^{-4}$$

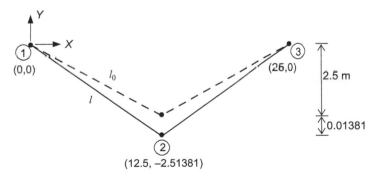

FIGURE 5.24e
Nodal coordinates after first load increment.

Force in the element $F = \sigma A = \varepsilon E A = 2.1965 \times 10^{-4} \times 12000 = 2.5563$ kN

$$[k_{E1}] = \begin{bmatrix} 941.157 & 0 & -941.157 & 0 \\ 0 & 0 & 0 & 0 \\ -941.157 & 0 & 941.157 & 0 \\ 0 & 0 & 0 & 0 \end{bmatrix}$$

$$[k_{G1}] = \begin{bmatrix} 0 & 0 & 0 & 0 \\ 0 & 0.20049 & 0 & -0.20049 \\ 0 & 0 & 0 & 0 \\ 0 & -0.20049 & 0 & 0.20049 \end{bmatrix}$$

$$[k_{t1}^1] = [k_{t1}^2] = [k_{E1}] + [k_{G1}]$$

For element 1: $c = 0.980372$, $s = -0.197157$
element 2: $c = 0.980372$, $s = 0.197157$

$$[K_{t1}^1] = \begin{bmatrix} 904.581 & -181.875 & -904.581 & 181.875 \\ -181.875 & 36.7765 & 181.875 & -36.7765 \\ -904.581 & 181.875 & 904.581 & -181.875 \\ 181.875 & -36.7765 & -181.875 & 36.7765 \end{bmatrix}$$

$$[K_{t1}^2] = \begin{bmatrix} 904.581 & 181.875 & -904.581 & -181.875 \\ 181.875 & 36.7765 & -181.875 & -36.7765 \\ -904.581 & -181.875 & 904.581 & 181.875 \\ -181.875 & -36.7765 & 181.875 & 36.7765 \end{bmatrix}$$

$$[K_{t1}] = \begin{bmatrix} 1809.162 & 0 \\ 0 & 73.55294 \end{bmatrix}$$

Load increment 2

$$\{\Delta P_{A2}\} = \begin{Bmatrix} 0 \\ -1 \end{Bmatrix} \text{kN}$$

$$\{\Delta D_{A2}\} = [K_{t1}]^{-1}\{\Delta P_{A2}\} = \begin{Bmatrix} 0 \\ -0.0136 \end{Bmatrix} \text{m}$$

At the end of second load increment,

$$\{P_{A2}\} = \{P_{A1}\} + \{\Delta P_{A2}\} = \begin{Bmatrix} 0 \\ -2 \end{Bmatrix} \text{kN}$$

$$\{D_{A2}\} = \{D_{A1}\} + \{\Delta D_{A2}\} = \begin{Bmatrix} 0 \\ -0.02741 \end{Bmatrix} \text{m}$$

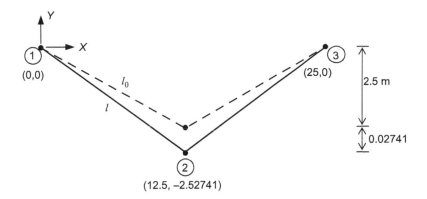

FIGURE 5.24f
Nodal coordinates after second increment.

The new nodal coordinates are shown in Figure 5.24f.

$$l = 12.75295 \text{ m}$$

$$F = 5.0861 \text{ kN}$$

For element 1: $c = 0.980165$, $s = -0.198182$

element 2: $c = 0.980165$, $s = 0.198182$

$$[K_{t2}] = \begin{bmatrix} 1808.034 & 0 \\ 0 & 74.68069 \end{bmatrix}$$

Load increment 3

$$\{\Delta P_{A3}\} = \begin{Bmatrix} 0 \\ -1 \end{Bmatrix} \text{ kN}$$

$$\{\Delta D_{A3}\} = [K_{t2}]^{-1} \{\Delta P_{A3}\} = \begin{Bmatrix} 0 \\ -0.01339 \end{Bmatrix}$$

$$\therefore \{D_{A3}\} = \{D_{A2}\} + \{\Delta D_{A3}\} = \begin{Bmatrix} 0 \\ -0.0408 \end{Bmatrix}$$

The above steps are repeated and the end of load increment = 100;

$$\{P_{A100}\} = \begin{Bmatrix} 0 \\ -100 \end{Bmatrix} \text{ kN}$$

$$\{D_{A100}\} = \begin{Bmatrix} 0 \\ -0.8939 \end{Bmatrix} \text{ m}$$

Method 2: Iterative Procedure Using N–R Method
Iteration 1

$$\{P_A\} = \left\{ \begin{array}{c} 0 \\ -100 \end{array} \right\} kN$$

$$\{\Delta P_{A1}\} = \{P_A\} = \left\{ \begin{array}{c} 0 \\ -100 \end{array} \right\} kN$$

$$[K_{t0}] = \left[\begin{array}{cc} 1810.303 & 0 \\ 0 & 72.41211 \end{array} \right]$$

$$\{\Delta D_{A1}\} = [K_{t0}]^{-1}\{\Delta P_{A1}\} = \left\{ \begin{array}{c} 0 \\ -1.38098 \end{array} \right\} m$$

$$\{D_{A1}\} = \{\Delta D_{A1}\} = \left\{ \begin{array}{c} 0 \\ -1.38098 \end{array} \right\} m$$

At the end of first iteration, the new coordinates of the nodes are shown in Figure 5.24g.

$$l = 13.0862 \text{ m}$$

$$\varepsilon = \frac{l-l_0}{l_0} = \frac{13.0862 - 12.7475}{12.7475} = 0.02676$$

$$F = \varepsilon AE = 321.0723 \text{ kN}$$

$$[k_{E1}] = \left[\begin{array}{cccc} 916.827 & 0 & -916.827 & 0 \\ 0 & 0 & 0 & 0 \\ -916.827 & 0 & 916.827 & 0 \\ 0 & 0 & 0 & 0 \end{array} \right]$$

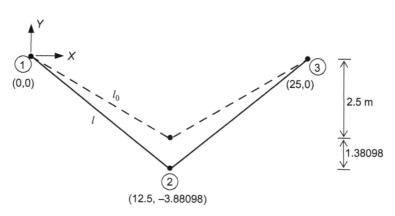

FIGURE 5.24g
Nodal coordinates after first iteration.

$$[k_{G1}] = \begin{bmatrix} 0 & 0 & 0 & 0 \\ 0 & 24.530 & 0 & -24.5306 \\ 0 & 0 & 0 & 0 \\ 0 & -24.5306 & 0 & 24.5306 \end{bmatrix}$$

$$[k_{t1}^1] = [k_{t1}^2] = [k_{E1}] + [k_{G1}]$$

For element 1: $c = 0.955028$, $s = -0.29652$

element 2: $c = 0.955028$, $s = 0.29652$

$$[K_{t1}^1] = \begin{bmatrix} 838.375 & -252.681 & -838.375 & 252.681 \\ -252.681 & 102.983 & 252.681 & -102.983 \\ -838.375 & 252.681 & 838.375 & -252.681 \\ 252.681 & -102.983 & -252.681 & 102.983 \end{bmatrix}$$

$$[K_{t1}^2] = \begin{bmatrix} 838.375 & 252.681 & -838.375 & -252.681 \\ 252.681 & 102.983 & -252.681 & -102.983 \\ -838.375 & -252.681 & 838.375 & 252.681 \\ -252.681 & -102.983 & 252.681 & 102.983 \end{bmatrix}$$

$$[K_{t1}] = \begin{bmatrix} 1676.749 & 0 \\ 0 & 205.9656 \end{bmatrix}$$

The FBD of node 2 is shown in Figure 5.24h.

In the figure, $F = 321.0723$ kN and $\sin\theta = 0.296516$. Hence,

$$P_e = 2F \sin\theta = 190.406 \text{ kN}$$

$$\therefore \{P_{e1}\} = \begin{Bmatrix} 0 \\ -190.406 \end{Bmatrix} \text{ kN}$$

Iteration 2

$$\{\Delta P_{A2}\} = \{P_A\} - \{P_{e1}\} = \begin{Bmatrix} 0 \\ 90.406 \end{Bmatrix} \text{ kN}$$

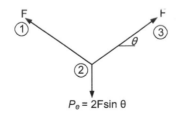

FIGURE 5.24h
FBD of node 2.

$$\{\Delta D_{A2}\} = [K_{t1}]^{-1}\{\Delta P_{A2}\} = \left\{ \begin{array}{c} 0 \\ 0.4389 \end{array} \right\} m$$

$$\{D_{A2}\} = \{D_{A1}\} + \{\Delta D_{A2}\} = \left\{ \begin{array}{c} 0 \\ -0.94208 \end{array} \right\} m$$

The coordinates of the nodes after the second iteration are shown in Figure 5.24i.

$$l = 12.96525 \text{ m, } F = 204.93385 \text{ kN}$$

For element 1: $c = 0.964116$, $s = -0.26548$
element 2: $c = 0.964116$, $s = 0.26548$

$$[K_{t2}] = \left[\begin{array}{cc} 1722.863 & 0 \\ 0 & 159.8521 \end{array} \right]$$

$$\{P_{e2}\} = \left\{ \begin{array}{c} 0 \\ -108.813 \end{array} \right\} kN$$

Iteration 3

$$\{\Delta P_{A3}\} = \{P_A\} - \{P_{e2}\} = \left\{ \begin{array}{c} 0 \\ 8.813 \end{array} \right\} kN$$

$$\{\Delta D_{A3}\} = [K_{t2}]^{-1}\{\Delta P_{A3}\} = \left\{ \begin{array}{c} 0 \\ 0.05513 \end{array} \right\} m$$

$$\{D_{A3}\} = \{D_{A2}\} + \{\Delta D_{A3}\} = \left\{ \begin{array}{c} 0 \\ -0.88695 \end{array} \right\} m$$

The coordinates of the nodes after third iteration are shown in Figure 5.24j.

$$l = 12.9507 \text{ m, } F = 191.2587 \text{ kN}$$

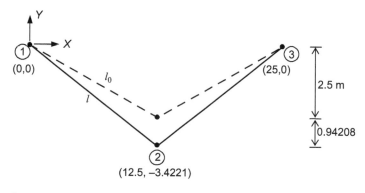

FIGURE 5.24i
Nodal coordinates after second iteration.

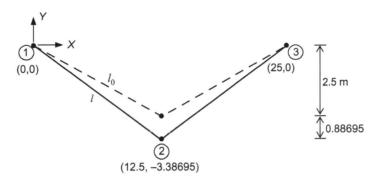

FIGURE 5.24j
Nodal coordinates after third iteration.

For element 1: $c = 0.9652$, $s = -0.26152$
element 2: $c = 0.9652$, $s = 0.26152$

$$[K_{t3}] = \begin{bmatrix} 1728.5 & 0 \\ 0 & 154.3 \end{bmatrix}$$

$$\{P_{e3}\} = \left\{ \begin{array}{c} 0 \\ -100.0372 \end{array} \right\} kN$$

Iteration 4

$$\{\Delta P_{A4}\} = \{P_A\} - \{P_{e3}\} = \left\{ \begin{array}{c} 0 \\ 0.0372 \end{array} \right\} kN$$

$$\{\Delta D_{A4}\} = [K_{t3}]^{-1} \{\Delta P_{A4}\} = \left\{ \begin{array}{c} 0 \\ 0.2414 \times 10^{-3} \end{array} \right\} m$$

$$\{D_{A4}\} = \{D_{A3}\} + \{\Delta D_{A4}\} = \left\{ \begin{array}{c} 0 \\ -0.8867 \end{array} \right\} m$$

The new coordinates of the nodes after third iteration are shown in Figure 5.24k.

$$l = 19.507 \text{ m}, F = 191.1993 \text{ kN}$$

$$\{P_{c4}\} = \left\{ \begin{array}{c} 0 \\ -99.9995 \end{array} \right\} kN$$

Since the difference between the total load and the equilibrated load is very small, the iteration is stopped at this stage. The final displacement obtained using the iterative procedure is

$$\{D_A\} = \left\{ \begin{array}{c} 0 \\ -0.8867 \end{array} \right\} m$$

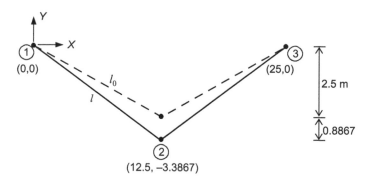

FIGURE 5.24k
Nodal coordinates after fourth iteration.

Problem

5.1 Analyze the following structures using the direct stiffness method.

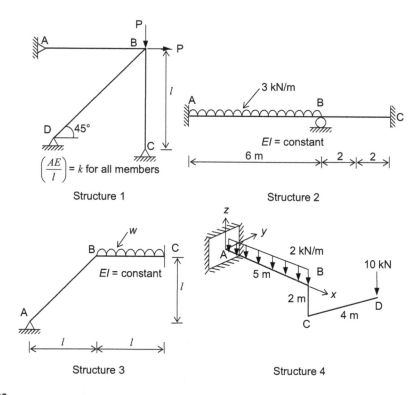

FIGURE 5.25
Problem 5.1.

Appendix

A sample MATLAB® code for the analysis of plane truss using the direct stiffness method is given below.

MATLAB code

```
% ANALYSIS OF PLANE TRUSS USING DIRECT STIFFNESS METHOD

% ne: Number of elements
% n: Number of nodes
% P: Force vector in global coordinates
% K: Global stiffness matrix
ne=input('Number of elements: ');
n=input('Number of nodes: ');
P=zeros(2*n,1);
K=zeros(2*n,2*n);

% Enter coordinate of nodes with respect to global axis
% coordinate(X-coordinate,Y-coordinate)
for i=1:n
    disp(['Enter Coordinates of node ',num2str(i)]);
    coordinate(i,1)=input('X Coordinate: ');
    coordinate(i,2)=input('Y Coordinate: ');
end

% Enter the degrees of freedom of the nodes i.e, global coordinates of
the nodes
% dof(horizontal dof, vertical dof)
disp('Enter numbering of degrees of freedom');
for i=1:n
    dof(i,1)=input(['Number of: Horizontal degree of freedom at node
',num2str(i),': ']);
    dof(i,2)=input(['Number of: Vertical degree of freedom at node
',num2str(i),': ']);
end

% Enter area and modulus of elasticity of the elements (A & E)
A=input('Area of element: ');
E=input('Modulus of Elasticity of element: ');
for i=1:ne
% Enter Near node and Far node of all elements
% node(Near node,Far node)
    disp(['Enter Properties of element ',num2str(i)]);
    node(i,1)=input('Node 1: ');
```

```
    node(i,2)=input('Node 2: ');

    % x: XF-XN
    % y: YF-YN
    % L: length of element
    % c, s: cos(theta), sin(theta)
    x=coordinate(node(i,2),1)-coordinate(node(i,1),1);
    y=coordinate(node(i,2),2)-coordinate(node(i,1),2);
    L(i)=sqrt((x^2)+(y^2));
    c(i)=x/L(i);
    s(i)=y/L(i);

    % EK: Stiffness matrix of one element
    EK=(A*E/L(i))*[c(i)*c(i) c(i)*s(i) -c(i)*c(i) -c(i)*s(i);
c(i)*s(i) s(i)*s(i) -c(i)*s(i) -s(i)*s(i);
-c(i)*c(i) -c(i)*s(i) c(i)*c(i) c(i)*s(i);
-c(i)*s(i) -s(i)*s(i) c(i)*s(i) s(i)*s(i)];

    % num: Global degrees of freedom of the element
    % num(horizontal dof at Near node,vertical dof at Near node
    %       horizontal dof at Far node,vertical dof at Far node)

num=[dof(node(i,1),1);dof(node(i,1),2);dof(node(i,2),1);dof(nod
e(i,2),2)];

    % Assembling stiffness matrix of the element to get the
    % Global stiffness matrix, K
    K(num,num)=K(num,num)+EK;
end

% restrained: Number of restrained degrees of freedom

% active: Number of active degrees of freedom
disp('Enter details of supports');
restrained=0;
i=1;
j=1;
while i==1
    % ns: Node number of support
    ns(j)=input('Node number: ');
    % sc: Support condition of the node
    % 1- Hinged, 2- Roller
    sc(j)=input('Support Condition: 1-Hinged; 2-Roller : ');
    if sc(j)==1
        restrained=restrained+2;
else
        restrained=restrained+1;
end
    i=input('1-More Supports; 2-No more Supports : ');
    j=j+1;
end
disp('Enter details of loads');
i=1;
while i==1
```

```
    % nl: Number of the node where the load is applied
    nl=input('Node number: ');
    Load=input('Load (N) : ');
    theta=input('Inclination of load (Anticlockwise from +x) (degrees) :
');

    % Save the component of the loads in the Global force vector
    P(dof(nl,1))=Load*cosd(theta);
    P(dof(nl,2))=Load*sind(theta);
    i=input('1-More Loads; 2-No more Loads : ');
end
active=(2*n)-restrained;

% K=[KAA KAR;KRA KRR]
KAA=K(1:active,1:active);
KAR=K(1:active,active+1:(2*n));
KRA=K(active+1:(2*n),1:active);
KRR=K(active+1:(2*n),active+1:(2*n));

% P=[PA;PR]    [active;restrained]
% D=[DA;DR]    [active;restrained]

PA=P(1:active);
DR=zeros(restrained,1);
DA=inv(KAA)*(PA-(KAR*DR));
PR=(KRA*DA)+(KRR*DR);
% ANALYSIS FINISHED


P=[PA;PR];
D=[DA;DR];

% Display support reactions
% ns contains node number of supports
for i=1:size(ns,2)
    disp(['Vertical Force at node ',num2str(ns(i)),' :
',num2str(P(dof(ns(i),2)))]);
    if sc(i)==1
        disp(['Horizontal Force at node ',num2str(ns(i)),' :
',num2str(P(dof(ns(i),1)))]);
    end
end

% Magnification factor-min(L)/(10*max(D))
D1=D*min(L)/(10*max(D));

for i=1:ne
    % num: Global degrees of freedom of the element
    % num(horizontal dof at Near node,vertical dof at Near node
%       horizontal dof at Far node,vertical dof at Far node)
num=[dof(node(i,1),1);dof(node(i,1),2);dof(node(i,2),1);dof(nod
e(i,2),2)];

    % T: Transformation matrix
```

```
    T=[c(i) 0;s(i) 0;0 c(i);0 s(i)];

    % p: element forces
    p=(A*E/L(i))*[1 -1;-1 1]*(T')*D(num);
    if p(1)<=0
        disp(['Axial force in element ',num2str(i),' : ',num2str(-p(1)),
' N (Tension)']);

else
        disp(['Axial force in element ',num2str(i),' : ',num2str(p(1)),
            ' N (Compression)']);
    end

% plot deformed shape

line([coordinate(node(i,1),1),coordinate(node(i,2),1)],[coordinate(node
(i,1),2),coordinate(node(i,2),2)],'Color','k','Linestyle','--');

line([coordinate(node(i,1),1)+D1(dof(node(i,1),1)),coordinate(node(i,2),1)
+D1(dof(node(i,2),1))],[coordinate(node(i,1),2)+D1(dof(node(i,1),2)),coord
inate(node(i,2),2)+D1(dof(node(i,2),2))]);
    xlabel('X (m)');
    ylabel('Y (m)');
end
```

Analysis of Plane Truss

The program is used to analyze the plane truss shown in Figure a.1. The discretized structure, elements, nodes, and global coordinates are shown in Figures a.2 and a.3, respectively. The element connectivity data are given in Table a.1.

The input data for the truss and the output data from the program are given below (Figure a.4).

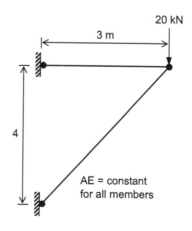

FIGURE a.1
Plane truss.

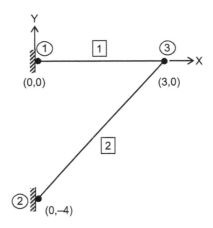

FIGURE a.2
Discretized structure.

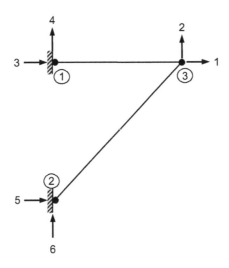

FIGURE a.3
Global coordinates.

TABLE a.1

Element Connectivity Data

Element	Near Node (N)	Far Node (F)
1	1	3
2	2	3

Input Data

Number of elements: 2
Number of nodes: 3
Enter Coordinates of node 1
X Coordinate: 0
Y Coordinate: 0
Enter Coordinates of node 2
X Coordinate: 0
Y Coordinate: –4
Enter Coordinates of node 3
X Coordinate: 3
Y Coordinate: 0
Enter numbering of degrees of freedom
Number of: Horizontal degree of freedom at node 1: 3
Number of: Vertical degree of freedom at node 1: 4
Number of: Horizontal degree of freedom at node 2: 5
Number of: Vertical degree of freedom at node 2: 6
Number of: Horizontal degree of freedom at node 3: 1
Number of: Vertical degree of freedom at node 3: 2
Area of element: 100
Modulus of Elasticity of element: 200000
Enter Properties of element 1
Node 1: 1
Node 2: 3
Enter Properties of element 2
Node 1: 2
Node 2: 3
Enter details of supports
Node number: 1
Support Condition: 1-Hinged; 2-Roller: 1
1-More Supports; 2-No more Supports: 1
Node number: 2
Support Condition: 1-Hinged; 2-Roller: 1
1-More Supports; 2-No more Supports: 2
Enter details of loads
Node number: 3
Load (N): 20

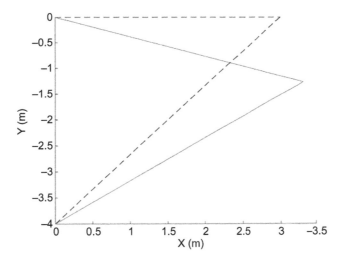

FIGURE a.4
Deflected shape of truss.

Inclination of load (Anticlockwise from +x) (degrees): 270
1-More Loads; 2-No more Loads: 2

Output

Vertical Force at node 1: 0
Horizontal Force at node 1: –15
Vertical Force at node 2: 20
Horizontal Force at node 2: 15
Axial force in element 1: 15 N (Tension)
Axial force in element 2: 25 N (Compression)

Bibliography

Arbabi, F., *Structural Analysis and Behaviour*, McGraw-Hill Education, New Delhi, 2014.

Balfour, J.A.D., *Computer Analysis of Structural Frame Works*, Second Edition, Oxford University Press, New York, 1992.

Beaufait, F.W., *Basic Concepts of Structural Analysis*, Prentice-Hall, New Jersey, 1977.

Beaufait, F.W., Rowan, W.H., Hoadley, P.G., and Hackett, R.M., *Computer Methods of Structural Analysis*, Prentice-Hall, New Jersey, 1970.

Chajes, A., *Structural Analysis*, Prentice-Hall, New Jersey, 1983.

Cook, R.D., Malkus, D.S., Plesha, M.E., and Witt, R.J., *Concepts and Applications of Finite Element Analysis*, Fourth Edition, John Wiley, Singapore, 2005.

Desai, C.S., and Abel, J.F., *Introduction to Finite Element Method*, CBS Publishers, Delhi, 2005.

Hibbeler, R.C., *Structural Analysis*, Fifth Edition, Pearson, New Delhi, 2002.

Hsieh, Y.Y., *Elementary Theory of Structures*, Third Edition, Prentice-Hall, New Jersey, 1988.

Jain, A.K., *Advanced Structural Analysis*, Nem Chand and Bros., Roorkee, 2006.

McGuire, W., Gallagher, R.H., and Ziemian, R.D., *Matrix Structural Analysis*, Second Edition, Wiley, New Delhi, 2000.

Meek, J.C., *Matrix Structural Analysis*, McGraw-Hill, New York, 1971.

Menon, D., *Advanced Structural Analysis*, Narosa, New Delhi, 2009.

Przemieniecki, J.S., *Theory of Matrix Structural Analysis*, Dover Publications, New York, 2012.

Rajasekaran, S., and Sankarasubramanian, G., *Computational Structural Mechanics*, Prentice-Hall of India, New Delhi, 2012.

Rao, S.S., *The Finite Element Method in Engineering*, Fifth Edition, Elsevier, New Delhi, 2013.

Rubinstein, M.F., *Matrix Computer Analysis of Structures*, Prentice-Hall, New Jersey, 1966.

Sennett, R.E., *Matrix Analysis of Structures*, Prentice-Hall, New Jersey, 1994.

Weaver, W., and Gere, J.M, *Matrix Analysis of Framed Structures*, Second Edition, CBS Publishers, Delhi, 1998.

Willems, N., and Lucas, W.M., *Structural Analysis for Engineers*, McGraw-Hill, Tokyo, 1978.

Index